同济大学研究生教材出版基金资助

地质工程专业英语

English for Geological Engineering

叶　斌　任非凡　编著

同济大学 出版社
TONGJI UNIVERSITY PRESS

图书在版编目(CIP)数据

地质工程专业英语 / 叶斌，任非凡编著. —上海：
同济大学出版社，2021.12
ISBN 978-7-5608-9511-6

Ⅰ.①地… Ⅱ.①叶… ②任… Ⅲ.①地质学—英语
—高等学校—教材 Ⅳ.①P5

中国版本图书馆 CIP 数据核字(2020)第 174714 号

地质工程专业英语
English for Geological Engineering
叶 斌 任非凡 **编著**

责任编辑 戴如月　　**助理编辑** 夏涵容　　**责任校对** 徐春莲　　**封面设计** 张　微

出版发行　**同济大学出版社**　　www.tongjipress.com.cn
　　　　　(地址:上海市四平路 1239 号　邮编:200092　电话:021-65985622)
经　　销　全国各地新华书店
排　　版　南京文脉图文设计制作有限公司
印　　刷　启东市人民印刷有限公司
开　　本　787 mm×1092 mm　1/16
印　　张　8.5
字　　数　212 000
版　　次　2021 年 12 月第 1 版　　2021 年 12 月第 1 次印刷
书　　号　ISBN 978-7-5608-9511-6

定　　价　58.00 元

前　言

　　研究生从事科学研究工作往往需要阅读大量的英文文献并撰写英文论文，因此掌握专业英语阅读和写作能力是研究生科研能力培养的一个重要环节。笔者从 2010 年起为同济大学地质工程专业的研究生讲授专业英语课程，在授课过程中深切感受到师生们需要一本专门针对本专业研究生的英语教材。于是笔者开始了相关材料的收集和教材的编写工作，经过一年左右的时间，编成了《地质工程专业英语》一书。

　　本书涉及地质学、地球物理、水文学、岩土力学、工程地质学等方面的专业知识，旨在培养地质工程专业研究生的英文文献阅读能力，并为其进一步提高专业论文的英文写作能力打下坚实的基础。

　　全书分为 5 个单元，每个单元包括 5 篇专业英文文献，全部取材于已发表的高质量期刊论文或专业教材，用语规范，针对性强。同时，为了便于读者使用，本书在英文原文之后均附上了词汇表和译文。

　　本书由同济大学叶斌和任非凡编著。研究生王宣望、冯晓青、马子骏、宋思聪、张旭东、张亮、冉艳霞、王晖皓、邹今航等同学参与了翻译和校对工作，在此表示衷心的感谢。

　　由于编者水平有限，本书如有不当之处，敬请读者批评指正。

编　者

2021 年 9 月

目　　录

Unit 1　Geology ·· 1

　　Text 1　A Unified Theory of the Earth ······································· 1

　　Text 2　Causes of Earthquakes and Lithospheric Plates Movement ··········· 6

　　Text 3　The Role of Bedding in the Formation of Fault-Fold Structure ······ 10

　　Text 4　Geological Map ·· 12

　　Text 5　Geological Development of an Area ································· 18

Unit 2　Geophysics ·· 23

　　Text 6　How Mountains Get Made ·· 23

　　Text 7　Seeing Is Believing ·· 27

　　Text 8　Taking Earth's Temperature ··· 32

　　Text 9　Magnetostatics ··· 37

　　Text 10　Geophysical Investigation Methods ······························· 42

Unit 3　Hydrology ··· 47

　　Text 11　A Decade of Sea Level Rise Slowed by Climate-Driven Hydrology

　　　　　　··· 47

　　Text 12　Ancient Geodynamics and Global-Scale Hydrology on Mars ········ 50

　　Text 13　Human-Induced Changes in the Hydrology of the Western United

　　　　　　States ·· 56

　　Text 14　Hydrological Cycle ··· 60

　　Text 15　Land Subsidence Due to Groundwater Drawdown in Shanghai ········ 65

Unit 4　Soil and Rock Mechanics ··· 71

　　Text 16　Acute Sensitivity of Landslide Rates to Initial Soil Porosity ········ 71

　　Text 17　Simple Scaling of Catastrophic Landslide Dynamics ················ 78

Text 18　Elements of Soil Physics ································· 82
Text 19　World Stress Map Project ······························· 87
Text 20　Can Buildings Be Made Earthquake-Safe?　·············· 92

Unit 5　Engineering Geology ··································· 98
Text 21　Assessing Landslide Hazards ························· 98
Text 22　Engineering Geological Assessment of the Obruk Dam Site ········ 103
Text 23　Engineering Geology — A Fifty Year Perspective ············· 110
Text 24　The Collapse of the Sella Zerbino Gravity Dam ············· 119
Text 25　Stability Analysis of Llerin Rockfill Dam ················ 124

References ·· 129

Unit 1 Geology

Text 1 A Unified Theory of the Earth

Because the presence of water is required for the formation of continental crust, the timing of continental crust implies that abundant surface water could not have been present early in the Earth's history. It is difficult to argue that the delayed formation of the crust can be the result of any other factor but the lack of water. For example, it may be possible that the early continental crust was destroyed by late, heavy bombardment. However, dating of cratering on the Moon indicates that the rate of bombardment by asteroids and meteorites had decreased to its present level by 3.85 Ga (Chyna and Sagan, 1992). Life was also present by 3.5 Ga; bombardments catastrophic enough to destroy continents likely would have sterilized the Earth. Another possibility is that water was present on the Earth's surface but did not enter the mantle. For example, if the mantle was not convecting before 3.0 Ga, then water would not have been transported into the mantle and little to no continental crust would have formed. However, convection is a mechanism of heat transport; its vigor is proportional to the thermal gradient. The Second Law of Thermodynamics requires that the Archean Earth would have been hotter than today, with higher thermal gradients. It is implausible that the Earth would cool for 1.5 billion years and only then begin mantle convection.

The scenario that is most consistent with the evidence is that most water accumulated on the Earth's surface post-accretion from a gradual — not sudden — process of extraterrestrial accretion. This is the only one of the four end-member theories of ocean formation that cannot be falsified; it is also the only theory that unifies and simplifies our understanding of the Earth. The theory accommodates both Ruby's (1951) arguments

1

for gradual accretion and the diverse modern evidence for an extraterrestrial source.

Two apparent difficulties remain. The time period from 4.5 to 3.85 Ga was a period of heavy bombardment by meteorites (Chyna and Sagan, 1992; Taylor, 1997), and logically the rate of water accretion would have also been highest early in the Earth's history. If rapid water accretion was occurring on a young Earth, why was the fastest rate of continental crust formation delayed until the late Archean? The second problem is the existence of Archean-age rocks that have been interpreted by some geologists as evidence that Archean oceans covered oceanic spreading ridges (e.g., Nesbit, 1987, p. 149). If Archean spreading ridges had been submerged, it might falsify a hypothesis of ocean origin by gradual accretion.

An answer to both of these objections is provided by a consideration of the role of topography. Probably the nearest modern analogue we have to the surface of the Early Archean world would be an ocean basin stripped of water. Although we do not know if the early Earth had seafloor spreading, lithospheric plates, and linear subduction zones, the young and hot Earth would almost certainly have had mantle convection. Isostasy requires substantial differences in elevation between young, hot oceanic lithosphere, and old, cold lithosphere. The elevation difference today between very young and very old oceanic lithosphere is about 3 km (Parsons and Slater, 1977). It is thus a logical and inescapable inference that under a scenario of gradual accumulation, water would have accumulated initially in the proto-subduction zones that occupied the lowest elevations. As the water volume increased over geologic time, the nascent oceans would have enlarged their areal coverage, but still been confined to the coldest areas. No doubt the topography was not monotonic, and local ponding and precipitation near hot areas would have resulted in some serpentinization of the exposed primitive crust. However, water would have had to cover the higher elevations of the ocean basins before large-scale hydration began. Large-scale hydration of oceanic crust did not likely take place until about 3.0 Ga, because the oceanic volume had not grown sufficiently to cover the young, hot lithosphere where hydrothermal circulation and serpentinization occur. This delay would explain why rapid growth of continental crust did not begin until around 3.0 Ga.

An ophiolite is a piece of hydrated oceanic crust that has been thrust up onto a continent. The general absence of ophiolites older than 2.5 Ga suggests that young oceanic lithosphere in the Early Archean did not undergo pervasive hydrothermal alteration and therefore was not covered by water. Greenstone belts contain sections that resemble parts of ophiolites, but the oldest known true ophiolite sequence has an age of 2.5 Ga (Musky et al., 2001).

Some Archean basalts and komatiites appear to have erupted subaqueously, implying that surface water was present on Earth from the earliest times. For example, komatiites from the Belingwe Belt in Zimbabwe formed pillow lavas. However，it is a mistake to interpret these rocks as evidence for the submergence of oceanic spreading ridges. Modern spreading ridges are not preserved，nor is it likely that their Archean analogues were. The only remnants of Archean age crust still in existence are those that were adjacent to，and caught up in，newly forming continents；today they are mostly found at the central core of continental cratons surrounded by rocks of younger age. Continental crust is created at subduction zones；therefore，those Archean rocks that are preserved today—including extrusive erupted under water—had to be generated near or in Archean proto-subduction zones. Even if the total amount of water present on the surface of the young Earth were some small fraction of the present volume，the water depth in these proto-subduction zones could have been several kilometers；the depth of modern ocean trenches commonly exceeds eight kilometers.

(Cited from Deming D. Origin of the ocean and continents：a unified theory of the earth ［J］. International Geology Review，2002，44(2)：137-152.)

New Words and Expressions

continental crust		大陆地壳
bombardment	n.	撞击
crater	v.	(月球上)成陨石坑
asteroid	n.	小行星
meteorite	n.	陨石
Ga	n.	(地质学时间单位)10亿年
catastrophic	adj.	大突变(灾难)的
sterilize	v.	使灭绝
mantle	n.	地幔
convect	v.	对流循环
Archean	adj.	太古代的
implausible	adj.	难以置信的
scenario	n.	(可能发生的)情况
accretion	n.	堆积
extraterrestrial	adj.	地球以外的
falsify	v.	证伪
interpret	v.	解释
ridge	n.	山脊
submerge	v.	使淹没

3

hypothesis	*n.*	假设
topography	*n.*	地貌
analogue	*n.*	相似物
basin	*n.*	盆地
strip	*v.*	剥除
lithospheric	*adj.*	岩石圈的
subduction	*n.*	俯冲
isostasy	*n.*	地壳均衡
inescapable	*adj.*	不可避免的
nascent	*adj.*	新生的
monotonic	*adj.*	单调的
precipitation	*n.*	降水
serpentinization	*n.*	蛇纹石化
primitive	*adj.*	原始的
hydration	*n.*	水化,水合(作用)
hydrothermal	*adj.*	水热的
circulation	*n.*	循环
ophiolite	*n.*	蛇绿岩
pervasive	*adj.*	遍布的
alteration	*n.*	变化,改变
basalt	*n.*	玄武岩
komatiite	*n.*	科马提岩
subaqueously	*adv.*	水下地
lava	*n.*	熔岩,岩浆
preserve	*v.*	保存,保留
remnant	*n.*	遗迹
craton	*n.*	(地)克拉通,稳定地块
extrusive	*n.*	喷出岩

译文:地球的统一理论

大陆壳的形成需要水,这意味着在大陆壳形成的地球历史早期不会有大量的地表水。人们很难认同大陆壳的延迟形成是由缺水以外的其他原因造成的。例如,有人认为早期的大陆壳可能被后来的大碰撞破坏了。但是对月球上陨石坑的形成年代的测定表明,来自小行星和陨石的撞击在38.5亿年前就已经下降到当前的水平(Chyna and Sagan,1992);生命也是在35亿年前出现,而能够破坏大陆壳的撞击则有可能使地球上的生命完全灭绝。另外有一种可能是水存在地球的表面而没有进入地幔。例如,如果地幔在30亿年前没有对流循环,水就不会被运送到地幔,这样就几乎没有大陆壳会形成。然而,对流是热传导的机制,它的活力与温度梯度成正比。根据热力学第二定律,太古时期的地球要比现在更热且有更高的温度梯度。因此,地球在冷却了15亿年后再开始地幔循环是不合常理的。

和现有证据最一致的情形是,大部分在地球表面的水来自地球以外的一个逐渐积聚(而不是突然增长)的过程。这是关于海洋形成的四个端元理论中唯一不能被证伪的理论,也是唯一能统一和简化我们对地球的理解的理论。这一理论也和Ruby(1951)关于水逐渐积聚的观点以及水来源于地球以外的各种现代证据相吻合。

这一理论也存在两个明显难以解释的疑点。从45亿年前到30亿年前是陨石大碰撞时期(Chyna and Sagan,1992;Taylor,1997),因此逻辑上水的积聚速度在地球历史早期是最高的。如果水的快速聚集发生在地球早期,那为什么大陆壳形成速度最快的时期却一直延迟到太古时代晚期?第二个问题是关于太古代的岩石,一些地质学家认为这些岩石的存在是太古代时期在扩张的海脊上有海洋覆盖的证据(e.g.,Nesbit,1987,p.149)。如果太古代扩张的海脊是沉没在海底的,这就否定了海洋是水逐步聚集而成的假设。

对这两个异议的一种解答是考虑地貌的作用。与太古代早期世界的地表形态最接近的现代相似物可能就是去除水以后的海洋盆地。虽然我们不知道地球早期是否存在海底扩张、岩石圈板块和线性俯冲区,但是年轻炽热的地球上肯定存在着地幔对流。地壳的均衡需要年轻炽热的海洋岩石圈和较老且较冷的岩石圈之间存在足够的高度差。今天最年轻和最古老的海洋岩石圈之间的高度差大概是3千米(Parsons and Slater,1977)。因此,一个自然且符合逻辑的推测就是,在水是逐渐积聚的情况下,它最初会在高程最低的原始俯冲区积聚。随着水的体积在地质历史时期逐渐增大,新生的海洋尽管也逐渐扩大覆盖面积,但是仍然限制在最冷的区域。当然,地貌并不是单调变化的,局部积水和降水可能会导致裸露的原始地壳蛇纹石化。然而,在大规模的水化作用发生之前,水必须覆盖海洋盆地的最高位置。海洋地壳的大规模水化作用直到30亿年前才发生,因为只有这时海洋的体积才能增长到足够覆盖年轻炽热且发生水热循环和蛇纹石化作用的岩石圈。这一延迟便可以解释为什么大陆壳的快速形成直到30亿年前才开始。

蛇绿岩是部分水化以后的海洋板块被挤入大陆板块后形成的。一般情况下不存在早于25亿年前形成的蛇绿岩,这意味着在太古代早期年轻的海洋岩石圈没有经历过普遍的热液蚀变,也没有被水覆盖。绿岩带中有部分类似于蛇绿岩,但是已知的最古老的蛇绿岩序列还是形成于25亿年前(Muskg et al.,2001)。

一些太古代的玄武岩和科马提岩似乎是在水下海底喷发形成的,这暗示在地球的最早时期有地表水存在。例如,在津巴布韦Belingwe带的科马提岩形成了枕状熔岩。然而,将这些岩石解释为海脊扩张沉没的证据是错误的。现代的扩张海脊并没有被保存下来,太古代的类似物也不可能被保存下来。现存的太古代地壳的唯一遗迹是那些毗邻新形成的大陆,并成为新大陆一部分的岩石。现今,它们主要位于被年轻的岩石包围的大陆稳定地块的中央从而得以被发现。大陆地壳是在俯冲带形成的,因此今天保存的那些太古代岩石,包括水下喷发的喷出岩,必然形成于太古代原俯冲带附近或太古代原俯冲带中。即使年轻的地球表面存在的水总量只是目前体积的一小部分,这些原俯冲带的水深也可达几公里;现代海沟的深度通常超过8公里。

Text 2 Causes of Earthquakes and Lithospheric Plates Movement

Consideration of the Earth's rotation as a factor influencing the Earth's surface is based on very old data. Already Darwin (1881) recognized that owing to the Earth's rotation, the equatorial regions are subjected to greater forces than the polar regions. Böhm von Böhmersheim (1910) presented an opinion that the Earth's rotation and its changes is an energy source of orogenetic processes. Next I mention authors Verona (1927), Schmidt (1948) and Stoves (1957). The development of Earth's rotation theories begins after the confirmation of Earth's rotation variations by comparison with the atomic clock and later by exact measurements using very long baseline interferometry (Munk and Mac-Donald, 1960). The detailed conception presented Chebanenko (1963), considering inertial forces acting on continents, requiring however the slip of the Earth's crust. Mac-Donald (1963) considers also the relation of the deep fault tectonics to rotation changes. After confirmation of the continental drift by interpretation of linear magnetic anomalies and dating of oceanic basalts by Morley and Vine and Mathews (1963) and introduction of the plate tectonics principles, hypotheses occurred considering tidal forces as driving agents of the plate movements (Bostrom, 1971; Knopoff and Leeds, 1972; Moore, 1973). These hypotheses were rejected by estimation of the mantle viscosity by Cathleen (1975), and Jordan (1975) presented a simple calculation that the mantle viscosity should be 10 orders of magnitude lower to make possible the movement, and most of geophysicists preferred the mantle convection as the plate driving agent originated in Holmes (1939) and later in McKenzie and Weiss (1975) and others. Later, Ranalli (2000) supported such hypotheses refusing the rotational drag as driving agent. Then 20 years ago at XX. General Assembly of the International Union of Geodesy and Geophysics in Vienna 1991, the author (Ostřihanský, 1991) presented a hypothesis that Earth's rotation variations trigger earthquakes and introduce lithospheric plates into movement. In the monograph "The causes of lithospheric plates movements" (Ostřihanský, 1997), the hypothesis was elaborated in detail. The two largest earthquakes of the beginning of this century, the large M7.9 Denali Fault Alaska earthquake in 2002 and the M9.1 Great Sumatra earthquake in 2004, confirmed this hypothesis (Ostřihanský, 2004). In the meantime, extensive investigations in global tectonic earthquakes have shown evidence of a correlation with diurnal tides (Tanaka et al., 2002; Cochran et al., 2004). The Sumatra 26 December 2004 earthquake was triggered not only on exact winter

Earth's rotation maximum speed but also on the full Moon. This inspired Crocket et al. (2006) to present a hypothesis of earthquake correlation with biweekly tides. This hypothesis was strictly refused by Cochran and Vidale (2007), presenting histograms of global data rejecting coincidence of earthquakes with biweekly tides. Supporters of the Earth's rotation effect are also Doglioni et al. (2007), showing that geological and geophysical asymmetries of rifts and subduction zones are a function of their polarity and may be interpreted as controlled by some astronomical mechanical shear (Scoppola et al., 2006). Crespi et al. (2007) have shown that plates follow a westward mainstream but inclined to equator $\pm 7°$. Riguzzi et al. (2009) presented paper summarizing previous results of Doglioni et al., Scoppola et al. and Crespi et al. Already for almost three decades ago two serious objections have been also raised against the triggering of earthquakes and the plate movement by Earth's rotation variations. The first one the Forsyth and Uyeda (1975) recognition that in oceans undoubtedly the ridge push and the slab pull act, disqualified the tidal forces acting in range 4×10^3 Pa for semidiurnal tides and 8×10^3 Pa for biweekly tides (Bodri and Iizuka, 1989). Ostřihanský (1997) however has shown that the slab-pull force represents the dropping down by gravity of the oceanic lithosphere opening the space for moving plates driven by week forces. The hydrostatic pressure (ridge-push) acts on the both sides, lower and upper, of the oceanic lithosphere, however this force acts directly in the oceanic ridge only in case when the drag following from the Earth's rotation opens the ridge and the ascending magma reaching the crest of the ridge acts by its pressure. This pressure accompanying pressures following from the Earth's rotation can be especially effective in case that the ridge is close to subduction zone. The second one the Wahr (1985) and Gipson and Ma (1998) calculations that LOD excites stress only 0.1 Pa seemingly excluded LOD. It was the same error as to consider the ITRF 2005 GPS measurement as the real lithospheric plates movements. These movements are related to the stable lithosphere to GPS satellite framework and this says nothing about the plate movements above mantle. In China the westward moving Eurasian plate collides with Indian plate moving in N-W direction. In that site Wang et al. (2000) measured the decadal (for 20 years) LOD correlated stress change in order of $10^4 - 10^5$ Pa. Wang et al. (2000) found this stress on tilt meters in western China where the moving westward Eurasian plate collides with the Indian plate. The purpose of this paper is to prove that just large continental plates with some oceanic parts are driven by LOD variations and from this reason the Wang's et al. (2000) measurement can be explained. The nature of the decadal LOD variations exhibiting irregular 60 years variations derived from stars occultations by Moon, available from the end of 18th century (Aoki et al., 1982), follows from distant (6 times weaker tidal forces than from the Moon) of large planets Jupiter, Saturn Uranus and Neptune (P. Kalenda, private communication，2012），which by their long time contribution to every Moon's and Sun's LOD variation，crate this irregular decadal variation. This phenomenon only underlines the importance of astronomical parameters for the Earth surface；however，the decadal variations in spite that ranging 3-4 ms，owing to 60 years time span，have no significant effect in earthquake triggering and the plate movement. Nevertheless，some indications show that in decadal LOD variations extremes，earthquakes are more numerous. This concerns for example the almost continuous number of earthquakes < 4 M in LOD minimum 2002-2006 (Ostřihanský，2010b) in Norcia-Marche-Abruzzi region of Apennines and the number of earthquakes in the wide LOD maximum 1964-1983 in Alaska described in Sect. 3.3.

(Cited from Ostřihanský L. Causes of earthquakes and lithospheric plates movement [J]. Solid Earth Discussions，2012，4：1411-1483.)

New Words and Expressions

lithospheric	*adj.*	岩石圈的
orogenetic	*adj.*	造山的
interferometry	*n.*	干涉法
inertial	*adj.*	惯性的
tectonic	*adj.*	构造的
continental drift		大陆漂移
hypotheses	*n.*	假说（hypothesis 的名词复数）
magnitude	*n.*	大小，量级
convection	*n.*	对流
diurnal tide		全日潮
biweekly tide		双周潮
histogram	*n.*	直方图
subduction zone		俯冲带
ascending	*adj.*	上升的
LOD		length of day 的简写，即"日长"

译文：地震与岩石圈板块运动的成因

将地球自转视为影响地表的一个因素是基于很早以前的数据：Darwin(1881)已经认识到，由于地球自转，赤道地区受到的力比极地地区大；Böhm von Böhmersheim(1910)提出了地球自转及其变化是造山过程的能量来源的观点；接着 Verona(1927)、Schmidt(1948)和

Stoves(1957)也提出了此观点。通过与原子钟的比较以及后来的利用长基线干涉测量法得到的精确测定,地球自转的变化得到了确认,地球自转的理论由此开始发展(Munk and Mac-Donald,1960)。Chebanenko(1963)提出了详细的概念,考虑到了作用于大陆的惯性力(然而这种力需要地壳的滑动)。Mac-Donald(1963)也考虑到了深断层构造与自转变化的关系。在 Morley、Vine 与 Mathews(1963)对线性磁异常的解释和大洋玄武岩的年代确定证实了大陆漂移说并引入了板块构造学原理之后,又出现了将潮汐力作为板块运动的驱动力的假说(Bostrom,1971;Knopoff and Leeds,1972;Moore,1973)。然而这些假说被 Cathleen(1975)对地幔黏度的估值否定。随后 Jordan(1975)给出了一个简单的计算,即地幔黏度只有低于 10 个数量级时才能够使板块产生运动。大多数地球物理学家更认可最早由 Holmes(1939),后来由 McKenzie 与 Weiss(1975)以及其他人提出的地幔对流是板块驱动力的观点。随后又有 Ranalli(2000)支持此类否认自转阻力为驱动力的假说。然后,20 年前在维也纳召开的 1991 年第 20 届国际大地测量学和地球物理学联合会大会上,作者(Ostřihanský,1991)提出了地球自转变化引发地震并使岩石圈板块运动的假说,并在专著《岩石圈板块运动成因》(Ostřihanský,1997)中详细阐述了该假说。本世纪初的两大地震,即 2002 年阿拉斯加迪纳利断层 7.9 级大地震和 2004 年苏门答腊 9.1 级大地震,证实了这一假说(Ostřihanský,2004)。与此同时,在全球构造地震的广泛调查中,已找到与全日潮存在关联的证据(Tanaka et al.,2002;Cochran et al.,2004)。2004 年 12 月 26 日苏门答腊地震不仅准确地发生在冬季地球达到最大自转速度时,而且当时还是满月。受这一现象启发,Crocket 等人(2006)提出了与双周潮相关的地震假说,但 Cochran 和 Vidale(2007)以充分的证据否定了该假说。他们给出了全球数据直方图,以此否定了地震与双周潮汐之间的巧合。地球自转效应的支持者还有 Doglioni 等人(2007),他们指出,裂谷和俯冲带的地质和地球物理的不对称性是一个极性函数,并且可以解释为受一些天文力学剪切的控制(Scoppola et al.,2006)。Crespi 等人(2007)曾经提出,板块向西流动,但向赤道倾斜 ±7°。Riguzzi 等人(2009)综述了 Doglioni 等人、Scoppola 等人以及 Crespi 等人之前的研究结果。在将近 30 年前,人们针对地球自转变化引发地震和板块运动的观点也提出了两个严肃的异议。第一个异议来自 Forsyth 和 Uyeda(1975)。他们认为,海脊的推力和板块的拉力作用无疑使作用于半日潮 $4×10^3$ Pa 范围内和作用于双周潮 $8×10^3$ Pa 范围内的潮汐力相抵消。然而,Ostřihanský(1997)却表示,板块拉力意味着受重力作用下沉的大洋岩石圈为受周力驱动的板块开辟了移动的空间。静水压力(洋脊推力)作用在大洋岩石圈的上下两侧,可是,这种力只有在地球自转产生的拖曳张裂了洋脊,以及上升的岩浆在压力作用下到达洋脊顶部的情况下才会直接作用于洋脊上。这种静水压力,伴随着来自地球自转的压力,在洋脊靠近俯冲带时会特别起作用。第二个异议来自 Wahr(1985)、Gipson 与 Ma(1998)。他们经过计算得出,日长激发应力仅为 0.1 Pa,看起来可以排除日长的影响。这一计算形式,与把 ITRF 2005 GPS 测量视为岩石圈板块的实际运动同样是错误的。这些运动与 GPS 卫星框架下的稳定岩石圈有关,但与地幔上方的板块运动无关。在中国,向西移动的欧亚板块与朝西北向移动的印度板块发生碰撞。Wang 等人(2000)在该位置经测量得到,与 20 年间日长相关的应力变化范围为 10^4 Pa 至 10^5 Pa。Wang 等人(2000)在中国西部向西移动的欧亚板块与印度板块碰撞的位置,借助倾斜仪发现了这种应力。本文旨在证明只有大的大陆板块和一些大洋板块是由日长的变化驱动的,并以此解释 Wang 等人(2000)的测量结果。从 18 世纪末开始,人们就可以利用月掩星得知 60 年不规则的日长年代际变化的本质(Aoki et al.,1982):它与木星、土星、天王星和海王星等大型行星的距离有关。虽然这些天体的距离对地球产生的潮汐力不足月球的 1/6(P. Kalenda,private communication,2012),但每天天体长时间对月球及太阳日长变化的累积作用最终导致了这种不规则的年代际变化。这一现象仅强调了天文参数对地球表面的重要性,然而,60 年间年代际的变化值为 3 至 4 ms,此结果对地震的引发和板块的运动不会产生显著的作用。尽管如此,一些迹象表明,日长年代际变化达到极值时,地震的次数更多。例如,从 2002 至 2006 年在亚平宁山脉 Norcia-Marche-Abruzzi 地区的最小日长值下小于 4 级的连续的地震次数和文章 3.3 小节中描述的从 1964 年至 1983 年在阿拉斯加的最大日长值下的地震次数就体现了这一规律性。

Text 3   The Role of Bedding in the Formation of Fault-Fold Structure

Faulting and folding are intimately related and represent the effects of combined brittle and ductile processes during deformation (Strayer and Hudleston, 1997). There are a wide variety of folds associated with faults (Ramsay and Huber, 1987; Price and Cosgrove, 1990; Butler, 1992). The geometry of these folds is controlled by the orientation and shape of the associated faults (Davis and Reynolds, 1996). Folds formed as a consequence of faulting include: fault-bend folds (Suppe, 1985), due to flexure of the hanging-wall as it is displaced over a thrust ramp, and fault propagation folds (Suppe, 1985), in which the fold geometry is determined by the fault shape and the accommodation of the loss of displacement at the fault tip of a thrust ramp. Folds formed earlier than the thrusts, referred to as break-thrust folds (Fischer et al., 1992; Gutiérrez Alonso and Gross, 1999) are related to high-angle faulting of the limb of a fold. Complex associations of folds and faults can also develop in imbricate fans or duplexes as a result of the progressive collapse of either the hanging-wall or the footwall (Boyer and Elliott, 1982). Deformation of imbricate fault systems by displacements along stepped surfaces is heterogeneous in space as a result of fault-fold interactions at different scales (Chester, 2003). The models of thrust sheet movement over stepped and curved fault surfaces, leading to the development of ramprelated folds, consider the fault geometry and the mechanical properties of the bedding as major influences on the kinematics of deformation (Strayer and Hudleston, 1997).

This paper reports an example of fault-fold interaction during transpression that affected a Lower Ordovician stratigraphic sequence within the Portalegre-Esperanc, a Shear Zone. A detailed macro- to meso-scale description of the fault-fold structures is used to explain how the competence contrast influences the geometry and nature of the resulting structures in a sequence of quartzites, slates and quartzo-feldspathic volcaniclastic rocks intruded by granites. The term competence is here used to denote the relative contrast in the viscosity of different rock layers within the deformed sequence. For example, layers of quartzites deformed to produce folds and faulting whilst the slates (less competent) and quartzo-feldspathic rocks deformed in a ductile manner, and occasionally accommodated strain by flowing between the quartzite beds. Textural studies were performed with the purpose of identifying the deformation mechanisms and to determine how strain localization in the stratigraphic sequence controlled fault nucleation. An additional issue that we explore is the relationship between the tectonic transport direction along faults and the associated fold vergence in transpressional shear zones. In most of the studies of zones deformed by thrust faults, reverse faults and folds are typically associated with a transport direction oriented at a high angle to the regional strike. However, in transpressional shear zones folding and faulting reveal a systematic subparallelism or low-angle relationship between the trend of tectonic transport direction (defined by the stretching lineation) and the trend of oblique-slip and/or strike-slip faults, fold axes and fold vergence.

(Cited from Pereira M F, Silva J B, Ribeiro C. The role of bedding in the formation of fault-fold structures, Portalegre-Esperança transpressional shear zone, SW Iberia [J]. Geological Journal, 2010, 45 (5-6): 521-535.)

New Words and Expressions

ductile	*adj*.	有韧性的,可延展的
orientation	*n*.	方向
flexure	*n*.	弯曲
ramp	*n*.	(地)断坡
propagation	*n*.	传播
imbricate fan		(地)叠瓦扇
Lower Ordovician		(地)下奥陶统
heterogeneous	*adj*.	各种各样的,成分混杂的
feldspathic	*adj*.	含长石的
denote	*v*.	表示,指示

译文:层理在断层—褶皱构造形成中的作用

断层和褶皱作用是紧密相关的,它们代表了变形过程中脆性和韧性过程相结合的影响(Strayer and Hudleston,1997)。有各种各样与断层相关的褶皱(Ramsay and Huber,1987;Price and Cosgrove,1990;Butler,1992),其几何形状是由相关断层的方向和形状控制的(Davis and Reynolds,1996)。由断层作用形成的褶皱包括断层上盘受逆冲作用沿断坡移动发生弯曲而形成的断层转折褶皱(Suppe,1985),以及断层传播褶皱,该褶皱的几何形状由断层形状和断层端部受逆冲作用沿断面位移逐渐损失的适应性决定(Suppe,1985)。早于逆冲前所形成的褶皱被称为断冲褶皱(Fischer et al.,1992;Gutiérrez Alonso and Gross,1999),其与褶皱翼部的高角度断裂有关。褶皱和断层的复杂联系也能在上盘或下盘连续性坍塌形成的叠瓦扇或双重构造中看到(Boyer and Elliott,1982)。由于不同尺度的断层与褶皱的相互作用,叠瓦断层系统沿台阶面的位移变形在空间上是不均匀的(Chester,2003)。导致与断坡相关褶皱发育的阶梯式和弧形断层面上逆冲岩片运动模型,将断层的几何形状和层理的力学性质视为变形运动受到的主要影响(Strayer and Hudleston,1997)。

本文举了一个在转化挤压过程中断层与褶皱相互作用的例子,该作用影响了Portalegre-Esperanca剪切带内下奥陶统地层层序。通过对断层褶皱构造的宏观到中等尺度的详细描述,解释了在花岗岩侵入的石英岩、板岩和石英—长石火山岩序列中,能力对比是如何影响其构造的几何形状和性质的。术语"能力"在这里用来表示变形序列中不同岩层黏度的对比。例如,石英岩层变形产生褶皱和断层,而板岩(能力较弱)和石英—长石岩以韧性的方式产生变形,偶尔通过在石英岩层之间流动来调节应变。进行结构研究的目的在于识别变形机制并确定地层层序中的应变局部化如何控制断层成核。我们探讨的另一个问题是沿断层的构造运移方向与走滑挤压剪切带中伴生的褶皱倾向之间的关系。在大多数逆冲断层变形带研究中,逆断层和褶皱通常与运移方向有关,该运移方向与区域走向呈高角度的关系。然而,在走滑挤压剪切带中,褶皱和断层在构造运移方向(由拉伸线理所定义)的走向与斜滑和/或走滑断层、褶皱轴和褶皱倾向之间表现出一种系统的次平行或低角度关系。

Text 4   Geological Map

A variety of structural techniques have been described in previous chapters. In the main, the approach has been one of dissecting the geological map and examining its parts. The map is, however, more than the sum of these geometrical parts, and it remains to consider some of the more collective features.

Properly done, the map is an exceedingly important tool in geology. The graphical picture it gives of the location, configuration and orientation of the rock units of an area could be presented in no other way. Essential as the map is, however, it is not without limitations, and if it is to be of maximum use these limitations must be fully understood. The most important point to realize is that geological maps generally record both *observations* and *interpretation*. In part, the element of interpretation is due to a lack of time and complete exposure; it is almost never possible to examine all parts of an area. If a complete map is to be produced, this lack of observed continuity then requires interpolation between observation points and such interpolation is, to a greater or lesser degree, interpretive.

To distinguish between observation and interpretation several devices may be adopted. Most commonly, special symbols are used to identify several degrees of certainty in the location of lithologic contacts (Fig. 4-1); additional map symbols can be found in Compton (1985, p. 372). The choice of these symbols depends both on the ability to locate the boundaries in the field and on the scale of the map. A common rule of thumb is that a solid line is used if the contact is known and located to within twice the width of the line (Kupfer, 1966). Accordingly, a very thin, carefully drawn pencil line 0.1 mm wide covers 1 m on a 1 : 10 000 map and is appropriate for a contact known within 2 m on the ground, and a thin ink line 0.3 mm wide is appropriate for a contact known within 6 m. Clearly, the detail that can be shown on a map is scale dependent.

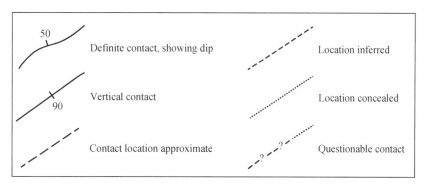

Fig. 4-1 **Map symbols for lithologic contacts**

If contacts are represented by less certain lines, it is important that their inferred locations make geometrical sense; if a contact line crosses a valley it should obey the Rule of Vs according to the inferred attitude. It is quite misleading to show uncertain contacts as if they were all vertical, though many examples of this practice can be found.

Factual and interpretive data may also be distinguished by considering the two aspects more or less separately. An outcrop map is one method of presenting field observations in a more objective way (Fig. 4-2a). Another way of conveying the essential information of an outcrop map, but without actually drawing in the boundaries of the exposed rock masses, is to show abundant attitude symbols, which then serve two functions: to record the measured attitude, and to mark the locality where the attitude can be measured.

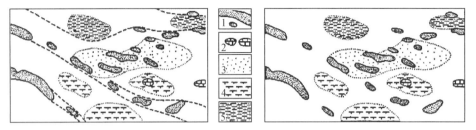

Fig. 4-2　Hypothetical maps. Lithologies: 1. sandstone; 2. limestone; 3. sandy soil; 4. limy soil; 5. clayey soil. (a) Outcrop and soil map showing facies interpretation (after Kupfer, 1966); (b) interpretation as a mélange; the blocks of sandstone and limestone are shown by outcrops and soils, the clay soil and covered areas are underlain by mélange matrix (after Hsu, 1968)

However, even an outcrop map or its equivalent can never be entirely objective for several reasons. What constitutes an exposure of rock is itself subject to some interpretation. For purposes of mapping, a thin rocky soil at the top of a low hill might be considered to be an outcrop by a worker in a poorly exposed area. In contrast, a geologist working in mountainous terrain would probably not give such an exposure a second glance because there are many better exposures. Such differences will certainly affect, and may even control the accuracy and completeness of the mapping.

Even with these limitations, it is, of course, important to strive for as high a level of objectivity as possible, and to discuss the difficulties involved in this quest in the text which accompanies the map.

There is another and much more fundamental reason why geological maps are inevitably interpretive. Even the simplest rock mass is extremely complex, and a complete physical and chemical description of a single outcrop could, quite literally, take years and questions concerning the origin of the rock would almost certainly remain. Clearly, such detailed studies are rarely feasible. The question then arises: What observations are to be made and recorded? The process of deciding what is important is guided in at least two ways. First, observations are made which have proven in the past to give results. Routine descriptions of attitude, lithology, visible structures and so forth are an important preliminary stage; some check lists have been published to facilitate this type of field description. However, the creative part of field study involves asking critical questions and then attempting to find the answers. These questions are formulated on the basis of knowledge, intuition and imagination. In this search for understanding, the older, often well-established approaches may actually be a barrier which must be broken through if progress is to be made.

For example, the interpretive aspect of the map of Fig. 4-2(a) is based on an application of the so-called laws of superposition, original horizontality, original continuity and faunal assemblage (Gilluly et al., 1968, pp. 92, 103). There are, however, rock bodies composed largely of sedimentary materials which do not obey these laws: a *mélange* is an example (Hsu, 1968). French for *mixture*, the term mélange is applied to a mappable body of deformed rocks consisting of a pervasively sheared, fine-grained, commonly pelitic matrix with inclusions of both native and exotic tectonic fragments, blocks or slabs which may be as much as several kilometers long (Dennis, 1967, p. 107). Fig. 4-2(b) is an interpretive map based on the recognition of a mélange.

Furthermore, the identification of even well-exposed rock is not always so straightforward that all geologist agree. And, as progress is made, concepts change. The most dramatic way of illustrating these changes is to compare two maps of the same area made at different times. One of the most startling examples, given by Harrison (1963, p. 228), involves a part of the Canadian Shield (Fig. 4-3). The earlier map was made at a time when "granites" were thought be entirely magmatic in origin. Later, a map was produced after it was realized that metamorphism and metasomatism could produce many of these same rocks. The result is that there is little in common between the two maps. This is an extreme example, but most geological maps still reflect, to a greater or lesser degree, the prejudices of the authors and their times.

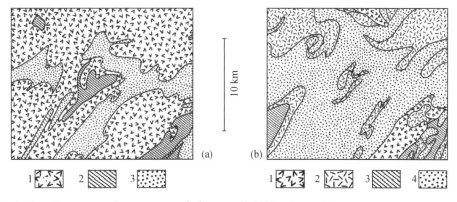

Fig. 4-3   Two maps of same area: (a) map of 1928: 1. granites with inclusions; 2. basic instructions; 3. metamorphic rocks with granite inclusions; (b) map of 1958: 1. granitic rocks; 2. migmatites with some granite; 3. basic intrusions; 4. metamorphic rocks (after Harrison, 1963)

As with most things, progress in mapping is an evolutionary process. Each step along the way is, at best, an approximation. These steps, because they are incomplete, necessarily involve some interpretation on the part of the investigator. Just as in the making of a geological map, so too does the use of the map as an aid to understanding the structure and history of an area involve several stages of development. The first step does not constitute making structural interpretations，but is rather a repetition of the experience and thinking of the original observer. This step is indispensable in gaining a complete understanding of both the map and the area it represents，and the facility to do this can only be achieved with practice. Two attitudes toward maps greatly increase their usefulness：

1. Regarding any geological map as a progress report. Improvement can always be made by further work based on the original mapping，either by the study of new exposures，or a more detailed study using new concepts and techniques.

2. Developing a critical outlook toward the lines and symbols on the map. By refusing to accept them completely，especially those that are clearly interpretive，and by adopting a questioning attitude toward the nature of the various map units and structures，new questions may arise that can be answered directly from the map，or from a visit to the area.

(Cited from Donal M R. Structural geology：An introduction to geometrical techniques [M]. 4th ed. Cambridge：Cambridge University Press，2009.)

New Words and Expressions

dissect	v.	剖析,解剖
geometrical	adj.	几何的
interpretive	adj.	解释的,作为说明的
attitude	n.	(地)产状;看法
outcrop	n.	露头
superposition	n.	叠加
politic	adj.	泥质的
exotic	adj.	外来的
metamorphism	n.	变质,变质作用
metasomatism	n.	交代作用

译文:地质图

在前面的章节中我们已经描述了各种结构方法,主要是对地质图进行剖析和部分检查。然而,地质图不仅仅是这些几何部分的总和,它仍然需要考虑一些更整体的特征。

正确完成的地质图是地质学中非常重要的工具。它给出的一个地区岩石单元的位置、构造和方位的图形描述是不能以其他方式呈现的。然而,虽然地质图必不可少,但它并非没有限制。如果要最大化地利用它,就必须充分了解这些限制。要意识到的最重要的一点是地质图通常同时记录"观察"和"解释"两部分。在某种程度上,地质图有关于解释的要素,是由于缺乏时间和完整的出露,几乎不可能检查一个区域的所有部分。如果要绘制一幅完整的地质图,这种观测的连续性的缺乏就需要在观测点之间进行插值,而这种插值或多或少是解释性的。

我们可以采用一些手段来区分观察和解释。最通常地,我们可以使用特殊符号来指明岩性接触位置的确定性程度的级别(图 4-1),其他的地质图符号可以在 Compton 所著的文献(1985,第 372 页)中找到。这些符号的选择既取决于定位范围内边界的能力,也取决于地质图的比例。一个常见的经验法则是,如果接触已知且位于线宽的两倍范围内(Kupfer,1966),则使用实线。相应地,一条非常细的、仔细绘制的 0.1 mm 宽的铅笔线相当于一幅比例为 1∶10 000 地图上的 1 m 的宽度,对地面 2 m 范围内已知的接触的呈现是合适的;而 0.3 mm宽的细墨水线则适合呈现地面 6 m 范围内已知的接触。显然,地质图上能够显示的细节的多少取决于比例。

若用较不确定的线表示接触,则其推断的位置应具有几何意义;如果一条接触线穿过山谷,则应按照 V 字法则来推断产状。将不确定的接触表示成垂直的会产生误导,尽管可以找到很多这种实际的例子。

真实的和解释性的数据也可以或多或少通过分别考虑这两个方面来区分。露头图是一种以更客观的方式呈现野外观测的方法[图 4-2(a)]。另一种表达露头图的基本信息而不用实际在裸露岩体的边界上绘制的方法,就是显示大量的产状符号,以起到记录被测产状和标定可测产状位置的作用。

然而,由于某些原因,即使是露头图或其等价物也不具有完全客观性,构成岩石出露的因素本身也有某些解释。为了绘制地质图,丘陵顶部的薄层岩质土可能被认为是在出露较差地区由劳动者活动产生的露头。相对应地,在山区工作的地质学家可能不会对这样的出露多看一眼,因为还有很多更好的出露。这样的差异肯定会有影响,甚至可能控制地质图的准确性与完整性。

当然,即使有这些限制,重要的是要争取达到尽可能高水准的客观性,并在地质图所附的文本中讨论所涉及的困难。

关于为什么地质地图具有不可避免的解释性,还有另外一个更根本的原因。即使是最简单的岩体也是极其复杂的,对单个露头进行完全的物理和化学描述可能确实需要数年时间,且有关岩石起源的疑问几乎肯定存在。显然,这种详细的研究很少可行。接下来的问题是:观察什么并将其记录下来? 至少有两种方法可以来指导决定哪些是重要的过程。首先,要进行观测,而这些观测在过去已经被证明是有结果的。对诸如产状、岩性和可见结构之类的常规描述是重要的初步阶段,为了便于这种类型的现场描述,已经发表过一些检查表。然而,现场研究最具创造性的部分包括关键问题的提出与随后的答案探寻。这些问题是建立在知识、直觉和想象的基础上的。在寻求理解的过程中,较陈旧的、通常已经确立的方法实际上可能是一个障碍。如果要取得进展,则必须突破这个障碍。

例如,地质图 4-2(a)解释的就是基于所谓的叠加定理、原始水平性、原始连续性和动物群落的应用(Gilluly, et al., 1968,第 92,103 页)。然而,岩体大部分是由不符合这些定律的沉积物质组成的。混杂岩（法文 *mélange*，英文 *mixture*）就是一个例子（Hsu，1968），其适用于长达数公里长的含有广泛剪切、细颗粒的和常夹杂有原生与外来构造碎屑、块体或板块的泥质基质的可绘制的变形岩体（Dennis，1967，第 107 页）。图 4-2(b)是基于混杂岩的解释图。

此外，所有地质学家都同意，即使是出露良好的岩石的鉴别也不总是如此简单。科学在进步，观念也随之改变。说明这些变化的最戏剧性的方法是比较两个在同一地区不同时间绘制的地质图。Harrison（1963，第 228 页）给出的最令人震惊的例子之一涉及加拿大地盾的一部分（图 4-3），其早期的地质图是在"花岗岩"被认为是完全起源于岩浆的时候绘制的。后来，在人们在认识到变质作用和交代作用能产生许多相同的岩石后，重新绘制了该地质图。结果这两张地图之间几乎没有共同点。这是一个极端的例子，但是大多数地质图仍然或多或少地反映了作者及其所处时代的偏见。

与大多数事物一样，绘图的进步是一个逐渐发展的过程。在这个过程中的每一步都是一个近似。这些步骤因为是不完整的，所以它们必然涉及对调查对象的部分解释。正如制作地质图一样，将地质图作为辅助工具去理解一个地区的结构和历史也涉及几个发展阶段。在第一步并不做出结构上解释，而是重复原始观察者的经验和思想。这个步骤对于获得对地质图及其代表区域的完整理解是必不可少的，并且只有通过实践才能实现。以下为提高地质图实用性的两点看法：

1. 把任何地质图作为进度报告。人们总是可以基于原始绘制的地质图开展进一步工作，或通过研究新的出露，或利用新的概念和方法来更详细地研究，对地质图进行改进。

2. 对地质图上的线条和符号进行批判性的展望。通过拒绝完全接受特别是那些解释清楚的地质图，以及通过对各种不同地质图单元和构造的性质持有质疑态度，可能会出现能够直接通过地质图或是访问该区域而得以解答的新问题。

Text 5　Geological Development of an Area

The geology of an area changes through time as rock units are deposited and inserted, and deformational processes change their shapes and locations.

Rock units are first emplaced either by deposition onto the surface or intrusion into the overlying rock. Deposition can occur when sediments settle onto the surface of the Earth and later lithify into sedimentary rock, or when as volcanic material such as volcanic ash or lava flows blanket the surface. Igneous intrusions such as batholiths, laccoliths, dikes, and sills, push upwards into the overlying rock, and crystallize as they intrude.

After the initial sequence of rocks has been deposited, the rock units can be deformed and/or metamorphosed. Deformation typically occurs as a result of horizontal shortening, horizontal extension, or side-to-side (strike-slip) motion. These structural regimes broadly relate to convergent boundaries, divergent boundaries, and transform boundaries, respectively, between tectonic plates.

When rock units are placed under horizontal compression, they shorten and become thicker. Because rock units, other than muds, do not significantly change in volume, this is accomplished in two primary ways: through faulting and folding. In the shallow crust, where brittle deformation can occur, thrust faults form, which causes deeper rock to move on top of shallower rock. Because deeper rock is often older, as noted by the principle of superposition, this can result in older rocks moving on top of younger ones. Movement along faults can result in folding, either because the faults are not planar or because rock layers are dragged along, forming drag folds as slip occurs along the fault. Deeper in the Earth, rocks behave plastically and fold instead of faulting. These folds can either be those where the material in the center of the fold buckles upwards, creating "antiforms", or where it buckles downwards, creating "synforms". If the tops of the rock units within the folds remain pointing upwards, they are called anticlines and synclines, respectively. If some of the units in the fold are facing downward, the structure is called an overturned anticline or syncline, and if all of the rock units are overturned or the correct up-direction is unknown, they are simply called by the most general terms, antiforms and synforms.

Even higher pressures and temperatures during horizontal shortening can cause both folding and metamorphism of the rocks. This metamorphism causes changes in the mineral composition of the rocks; creates a foliation, or planar surface, that is related to mineral growth under stress. This can remove signs of the original textures of the rocks, such as bedding in sedimentary rocks, flow features of lavas, and crystal patterns in crystalline rocks.

Extension causes the rock units as a whole to become longer and thinner. This is primarily accomplished through normal faulting and through the ductile stretching and thinning. Normal faults drop rock units that are higher below those that are lower. This typically results in younger units being placed below older units. Stretching of units can result in their thinning; in fact, there is a location within the Maria Fold and Thrust Belt in which the entire sedimentary sequence of the Grand Canyon can be seen over a length of less than a meter. Rocks at the depth to be ductilely stretched are often also metamorphosed. These stretched rocks can also pinch into lenses, known as boudins, after the French word for "sausage", because of their visual similarity.

Where rock units slide past one another, strike-slip faults develop in shallow regions, and become shear zones at deeper depths where the rocks deform ductilely. The addition of new rock units, both depositionally and intrusively, often occurs during deformation. Faulting and other deformational processes result in the creation of topographic gradients, causing material on the rock unit that is increasing in elevation to be eroded by hillslopes and channels. These sediments are deposited on the rock unit that is going down. Continual motion along the fault maintains the topographic gradient in spite of the movement of sediment, and continues to create accommodation space for the material to deposit. Deformational events are often also associated with volcanism and igneous activity. Volcanic ashes and lavas accumulate on the surface, and igneous intrusions enter from below. Dikes, long, planar igneous intrusions, enter along cracks, and therefore often form in large numbers in areas that are being actively deformed. This can result in the emplacement of dike swarms, such as those that are observable across the Canadian shield, or rings of dikes around the lava tube of a volcano.

All of these processes do not necessarily occur in a single environment, and do not necessarily occur in a single order. The Hawaiian Islands, for example, consist almost entirely of layered basaltic lava flows. The sedimentary sequences of the mid-continental United States and the Grand Canyon in the southwestern United States contain almost-undeformed stacks of sedimentary rocks that have remained in place since Cambrian time. Other areas are much more geologically complex. In the southwestern United States, sedimentary, volcanic, and intrusive rocks have been metamorphosed, faulted, foliated, and folded. Even older rocks, such as the Acasta gneiss of the Slave craton in northwestern Canada, the oldest known rock in the world have been metamorphosed to the point where their origin is undiscernable without laboratory analysis. In addition, these processes can occur in stages. In many places, the Grand Canyon in the southwestern United States being a very visible example, the lower rock units were metamorphosed and deformed, and then deformation ended and the upper, undeformed units were deposited. Although any amount of rock emplacement and rock deformation can occur, and they can occur any number of times, these concepts provide a guide to understanding the geological history of an area.

(Cited from Geological development of an area [EB/OL]. https://en.wikipedia.org/wiki/Geology.)

New Words and Expressions

deposit	v.	沉积
deformation	n.	变形
intrusion	n.	侵入
sediment	n.	沉积物
lithify	n.	岩化

lava	*n*.	岩浆
batholith	*n*.	岩基
dike	*n*.	岩墙
sill	*n*.	岩床
crystallize	*n*.	结晶
initial	*adj*.	最初的
metamorphose	*n*.	变质
convergent	*adj*.	收敛的
divergent	*adj*.	发散的
crust	*n*.	地壳
tectonic	*adj*.	构造的
superposition	*n*.	重叠,叠加
brittle	*adj*.	脆性的
planer	*adj*.	平坦的
antiform	*n*.	背形
synform	*n*.	向形
anticline	*n*.	背斜
synclines	*n*.	向斜
foliation	*n*.	叶理
intrusive	*n*.	侵入
topographic	*adj*.	地形学的

译文:一个区域的地质发展

随着岩石单元的沉积和侵入,一个地区的地质条件随时间而发生变化,并且这一过程改变了它们的形状和位置。

岩石单元首先通过沉积在地表或侵入上覆岩石而到达成岩的位置。当沉积物沉降到地表并岩化为沉积岩时,或当火山灰、熔岩流等火山物质覆盖地表时,就会发生沉积作用。火成岩侵入体,如岩基、岩盘、岩墙和岩床,向上进入上覆岩层中,并在侵入时结晶。

在岩石按照初始序列沉积后,岩石单元可以变形并/或发生变质作用。变形通常是水平收缩、水平拉伸或侧向(走向滑动)运动的结果。这些构造体制大致上与构造板块之间的聚合型边界、离散型边界和转换型边界有关。

当岩石体处于水平压缩状态时,它们变短、变厚。这主要通过两种方式实现:断层和褶皱。这是因为岩石单元不同于泥浆,体积不会发生显著变化。在可能发生脆性变形的浅部地壳中,会形成逆冲断层,使深部岩石在浅部岩石顶部移动。因为深层岩石通常较老,正如叠加原理指出的那样,这可能导致较老的岩石在更年轻的岩石上移动。沿着断层的运动可以导致褶皱,可能是因为断层不是平面的,也可能是因为岩石层被拖曳着,岩层沿着断层发生滑动时会形成拖曳褶皱。在地球深处,岩石表现为塑性和褶皱,而不是断层。这些褶皱可以是褶皱中央的物质向上弯曲,形成"背斜形态";也可以是向下弯曲,形成"向斜形态"。如果褶皱中岩石单元顶部保持指向上方,则被称为背斜或向斜;如果褶皱中的一些单元朝向下方,则称之为倒转背斜或倒转向斜;如果所有的岩石单元都被倒转或正确的向上方向未知,则简单地用最概括的术语——背斜形态和向斜形态来称呼它们。

水平收缩过程中压力和温度升高也会导致岩石的褶皱和变质。这种变质作用引起岩石矿物成分的变化,在应力作用下形成与矿物生长有关的片理或平面表面。这可以消除岩石原始纹理的痕迹,如沉积岩的层理、熔岩的流动特征、结晶岩的结晶形式等。

拉伸能使整个岩石单元变长、变薄。这主要是通过正断层作用及延性伸展和变薄来完成的。正断层会使原来较高的岩石单元移动到原来较低的岩石单元之下,这通常导致较年轻的岩石单元被置于较老单元之下。岩石单元的伸展会导致岩石单元变薄。事实上,在 Maria 褶皱和冲断带中一个位置上,可以在不到一米的长度上看到整个大峡谷的沉积序列。位于一定深度被延性拉伸的岩石也经常发生变质作用。这些被拉伸过的岩石也可能被挤压成透镜体形状且由于它们在视觉上的相似性,又被称为"boudins",这在法语中是"香肠"的意思。

岩体发生相互滑动的地方,在浅部会发育走滑断层,而在深部发生延性变形的地方会形成剪切带。新的岩石单元增加,无论是沉积的还是侵入的,常常发生在变形过程中。断层和其他变形过程会产生地形梯度,导致岩石单元上高度增加的物质被山坡和沟谷侵蚀。这些沉积物在正在下沉的岩石单元上沉积。沿着断层的持续运动保持了地形梯度而不受沉积物移动的影响,并继续为物质沉积创造容纳空间。变形活动也经常与火山作用和岩浆活动有关。火山灰和熔岩堆积在地表,岩浆侵入体从下面侵入。岩墙,作为长而平的火成侵入体,会沿着裂缝侵入,因此经常在活动变形区域大量形成。有可能形成岩墙群,比如加拿大地盾区的岩墙群,或者火山熔岩管周围的岩墙圈。

所有这些过程不一定发生在单一的环境中,也不一定以单一的顺序发生,例如,夏威夷群岛几乎完全由玄武质的熔岩流组成。美国中部大陆和西南部大峡谷的沉积层序中包含了自寒武纪以来一直存在的几乎未变形的沉积岩堆。其他地区的地质情况要复杂得多。在美国西南部,沉积、火山和侵入岩已经历了变质作用、断层作用、叶理作用和褶皱作用。更古老的岩石,如加拿大西北部 Slave 克拉通的阿斯塔片麻岩这种世界上已知的最古老的岩石,也已经变质到没有实验室分析就无法辨认的地步。此外,这些过程可以分阶段进行。在许多地方,例如美国西南部的大峡谷,下部岩石单元变质变形,然后变形结束,上部未变形单元沉积。虽然任何数量的岩石侵位和岩石变形都可能发生,而且可能发生多次,但这些概念为理解一个地区的地质历史提供了指引。

Unit 2　Geophysics

Text 6　How Mountains Get Made

The formation of mountain belts (orogens) in subduction-collision setting, where an oceanic plate subducts beneath continental material, is a fundamental process in plate tectonics. However, the mechanisms by which the continental crust deforms to produce shortening and uplift, and thus high topography, has been a matter of debate. This uncertainty is largely due to the difficulty of making direct observations of deformation in the deep crust to test the predictions made by conceptual models. On Page 720 of this issue, Huang et al. use observations of seismic anisotropy to constrain the geometry of deformation in the continental crust beneath the Taiwan orogen, and thus shed light on how the crust deforms as mountains are formed.

Two general concepts of crustal deformation in collisional orogens have been proposed, known as the thin-skinned and thick-skinned models. In the thin-skinned model, deformation is accommodated mainly in the upper crust, with a mechanically weak detachment surface (a décollement) separating the deforming upper layers from the deeper crustal rocks. In this scenario, the horizontal shortening and uplift required to form the mountains are confined to the upper crust. In contrast, the thick-skinned model invokes the deformation of the deeper crust as well as its shallow portions; here, the basement crustal rocks (as well as, perhaps, the mantle lithosphere beneath) undergo appreciable deformation. Taiwan represents an excellent locality to test these conceptual models; it is a young, actively deforming collisional orogen that accommodates the ongoing convergence between the Eurasian and Philippine Sea plates.

A key challenge in discriminating among the different models of crustal deformation is the difficulty of constraining deformation in the deep crust. One type of observation that can shed light on deformation in the deep Earth is the characterization of seismic anisotropy, or the directional dependence of seismic wave speeds. In many regions of the Earth, including much of the crust, there is a relationship between strain and the resulting anisotropy: As a rock deforms, individual mineral crystals tend to rotate and form a statistical preferred alignment, giving rise to seismic anisotropy. Therefore, observations of anisotropy in the crust can constrain the depth distribution of collision related deformation in orogens. Detailed observations of crustal anisotropy (and its variation in three dimensions) can be difficult, but recent innovations in observational seismology have advanced its study. These include the increasing availability of dense networks of seismometers, and the use of the ambient seismic noise field to extract information about crustal structure, including its anisotropy.

Huang et al. develop a tomographic model of shear wave velocity and anisotropy in the crust beneath Taiwan using measurements of surface waves derived from ambient noise. They find evidence for a sharp change in the geometry of seismic anisotropy at a depth of around 10 km to 15 km in the crust. Above this transition region, the fast directions of anisotropy are roughly parallel to the strike of the Taiwan orogen, and correlate closely with surface geologic trends. The authors propose that anisotropy in this upper layer is induced by compressional deformation and shortening. Beneath the transition, the fast directions of anisotropy are roughly parallel to the direction of convergence between the Eurasian and Philippine Sea plates. Here, the authors hypothesize that the deeper layer of anisotropy is caused by shear deformation of anisotropic minerals in the deep crust, induced by the motion of the down going plate.

As to what the results tell us about crustal deformation in the Taiwan orogen, and which of the thin-skinned or thick-skinned concepts apply, Huang et al. propose a hybrid model that has aspects of both concepts. The presence of strong anisotropy in the lower crust induced by convergence parallel shearing implies that there is deformation throughout the crust, as suggested by the thick-skinned model. On the other hand, the evidence for a sharp change in deformation geometry at a depth of 10 km to 15 km shares some aspects of the thin-skinned concept. The authors propose that this transition is not a décollement in the traditional sense, as there is mechanical coupling between the upper and lower crustal layers; however, their model does imply that compressional tectonics is active only in the upper layer.

These results reported by Huang et al. have important implications for our understanding of how the crust deforms in collisional orogens, and may prompt a reexamination of other mountain belts. A key question is whether there is widespread lower crustal anisotropy in other orogens, and whether a transition in deformation style in the mid-crust is a universal feature. The implications of such a sharp transition for our understanding of crustal rheology need to be explored. Another important question is to what extent the mantle lithosphere, in addition to the lower crust, participates in deformation. More generally, the observation and interpretation of crustal anisotropy, both in mountain belts and in other tectonic settings, represents an exciting frontier area, enabled by the increasing availability of data from dense seismic networks and the maturation of observational techniques that rely on the ambient noise field or on the analysis of converted waves. Furthermore, new constraints on the relationships between strain and anisotropy in crustal rocks are enhancing our ability to relate seismic observations to deformation geometry, opening the door to the detailed and quantitative testing of hypotheses related to the deformation of Earth's crust.

(Cited from Long M D. How mountains get made [J]. Science, 2015, 349(6249): 687-688.)

New Words and Expressions

orogens	n.	造山带
subduction	n.	俯冲
intrusion	n.	侵入
sediment	n.	沉积物
collision	n.	碰撞
continental	adj.	大陆性的
tectonics	n.	构造学
anisotropy	n.	各向异性
seismic	adj.	地震的
geometry	n.	几何构造
detachment	n.	分离,分隔
décollement	n.	滑脱构造
scenario	n.	设想,方案
discriminate	v.	识别,辨别
alignment	n.	队列
tectonic	adj.	构造的
dimensions	n.	规模,大小
tomographic	adj.	层析成像的
velocity	n.	速度
ambient	adj.	周围的,外界的
compressional	adj.	有压缩性的

frontier	*n*.	边界;前沿
dense	*adj*.	稠密的
hypothesis	*n*.	假定,猜想

译文:山脉是如何形成的

俯冲碰撞背景下的山带(造山带)的形成是板块构造的基本过程。在这一过程中,大洋板块俯冲至大陆板块下方。然而,大陆地壳变形产生缩短和抬升,从而形成高地貌的机制一直是一个有争议的话题。这种不确定性主要是因为人们难以直接观测地壳深部的变形以检验根据概念模型做出的预测。在本期的第720页,Huang等人利用对地震各向异性的观测来确定台湾造山带下方大陆地壳变形的几何结构,从而阐明地壳如何变形为山。

主要有两种概念来阐述碰撞造山带的地壳变形,即薄壁模型和厚壁模型。在薄壁模型中,变形主要存在于上地壳中,由一个力学上较弱的脱离面(滑脱构造)将变形的上层与较深的地壳岩石分开。在这种情况下,形成山脉所需的水平缩短和抬升仅限于上地壳。相反,厚壁模型则反映了深部及浅部地壳的变形。在这一模型中,基底地壳岩石(或许还有下面的地幔岩石圈)经历了明显的变形。台湾地区是检验这些概念模型的好地方:它是一个年轻的、变形活跃的碰撞造山带,在此欧亚板块和菲律宾板块之间正在进行着融合。

区分地壳变形的不同模式的一个关键挑战是难以确定深地壳变形。有一种观测地球深部变形的方法是利用地震各向异性的表征,或者地震波速度的方向依赖性。在地球的许多地区,包括地幔的大部分,应变和由此产生的各向异性之间存在着如下关系:随着岩石变形,单个矿物晶体倾向于旋转并形成统计上的优势排列,从而引起地震各向异性。因此,对地壳中各向异性的观测可以确定造山带碰撞相关变形的深度分布。详细观测地壳各向异性(及其三维变化)可能是困难的,但是最近观测地震学的创新使相关研究取得了进展。其中包括密集的地震计网络得到了越来越多的使用,以及利用环境地震噪声场提取有关地壳结构(包括其各向异性)的各种信息等。

Huang等人利用由环境噪声导出的表面波测量结果,建立了台湾地区地壳剪切波速度和各向异性的层析模型。他们发现了地壳中大约10千米至15千米深处地震各向异性的几何结构发生急剧变化的证据。在该转换区上方,各向异性的快方向大致平行于台湾造山带的走向,且与地表地质趋势密切相关。作者认为,该层的各向异性是由挤压变形和缩短引起的。在转换区之下,各向异性的快方向大致平行于欧亚板块和菲律宾板块之间的融合方向。这里,作者假设深部各向异性层是由深部地壳中各向异性矿物的剪切变形引起的,而这些剪切变形是由下沉板块的运动造成的。

关于台湾造山带地壳变形的结果,以及薄壁和厚壁哪个概念模型更为合适这个问题,Huang等人提出了一个兼具两个概念模型的某些特征的混合模型。下地壳存在的与融合方向平行的剪切引起的强各向异性,证明整个地壳存在变形,这与厚壁模型的观点一致;另一方面,10千米至15千米深度处变形几何结构的急剧变化的迹象与薄壁概念模型的某些方面相一致。作者提出,这种转换不是传统意义上的滑脱构造,因为上地壳层和下地壳层之间存在机械耦合;然而,他们的模型确实表明压缩构造仅在上层是活跃的。

Huang 等人的这些研究结果对我们理解碰撞造山带地壳如何变形有重要意义,并可以促使我们重新审视其他造山带。一个关键问题是,在其他造山带中是否普遍存在下部地壳各向异性,以及中部地壳变形方式的转变是否是一个普遍的特征。为了理解地壳流变学的意义,这种急剧转变需要更深入的探讨。另一个重要问题是,除了下部地壳以外,地幔岩石圈在多大程度上参与了变形。更一般地来说,在山区和其他构造环境中,地壳各向异性的观测和解释都代表了一个令人兴奋的前沿领域,这得益于来自密集地震网络的更多的可用数据和依赖于环境噪声场或转换波分析的观测技术的成熟。此外,对地壳岩石应变与各向异性之间关系的新发现正在增强我们将地震观测与变形几何学联系起来的能力,为详细和定量检验与地壳变形有关的假设打开大门。

Text 7   Seeing Is Believing

The lava lamp on my son's bureau gives him a vantage point to ponder the colorful blobs rising and falling as their temperature and density change. Geologists are not so privileged. Some of Earth's most impressive geologic features, such as the massive granitic plutons of California's Sierra Nevada mountains, are direct consequences of magma transport but we cannot directly observe these transport processes. On Page 250 of this issue, Fialko and Pearse use satellite data and computational modeling to infer the transport process in one location. They conclude that a magmatic diapir—a roughly spheroidal mass of partially molten material—is rising beneath a portion of the Altiplano Plateau in the central Andes.

To deduce subsurface processes, geologists must compare displacements of millimeters to centimeters per year on Earth's surface with mathematical models. However, different models can give similar predictions within the observational errors, and the most physically plausible models are often too computationally intensive to allow adequate comparison with the data. Fundamental advances thus require some combination of increased observational and computational power. In their tour-de-force study, Fialko and Pearse achieve both.

The authors used the interferometric synthetic aperture radar (InSAR) technique with data from three different satellites to create an 18-year record of surface displacement at Bolivia's Uturuncu volcano. The study region sits directly above what is believed to be the largest active magma body in Earth's continental crust: the Altiplano-Puna Ultralow-Velocity Zone (APULVZ), which is ~17 km to 19 km deep, ~1 km thick, and ~100 km wide.

Previous studies, based on 8 years of InSAR data from two satellites with similar viewing geometry, identified surface uplift associated with Uturuncu and modeled it as the manifestation of an inflating magmatic point source at a depth of ~15 km to 17 km. To better constrain the depth, geometry, and time history of the supposed magmatic source, Fialko and Pearse tasked a third satellite to acquire data from a very different look direction. Together with the previously acquired data, this more robust data set revealed an unexpected pattern of subsidence around the periphery of uplift.

The combined uplift and subsidence is difficult to explain with models that assume elastic behavior of Earth's crust. Fialko and Pearse therefore tested increasingly sophisticated viscoelastic models. Their compelling conclusion is that the APULVZ is a magma source for a diapir that rises toward the surface. The rising, ballooning diapir causes the surface uplift; simultaneous withdrawal of material from the APULVZ causes the peripheral subsidence.

It is easy to agree that diapirism occurs in a lava lamp: We can see it happening. But Earth's crust consists of rock with much higher viscosity; there has been no consensus on whether diapirism is a viable magma-transport process. Objections have centered mostly on the lack of field evidence for rocks intensely deformed by passing diapirs.

These objections caused many to argue that silicic magmas are transported in bladelike dikes, of which there are myriad field examples, albeit with low-viscosity iron- and magnesium-rich compositions. On the other hand, more recent thermomechanical analysis suggests that we should not discard the possibility of diapirs, because dikes with more viscous, silicic compositions may freeze within a few hundred meters from their source region.

Fialko and Pearse's study provides the strongest evidence to date for the existence of diapirs. The obvious next question is: How prevalent are they? InSAR studies have detected many inflation signals associated with volcanoes around the world, especially in the Andes (see Fig. 7-1). If each of those centers were studied as thoroughly as Uturuncu, how many would turn out to be diapir-fed? A more time-critical question, however, is how the new findings affect hazards. A large and rising magma body sounds ominous. Uturuncu has not erupted for ~270,000 years, but the region has experienced massive eruptions during the past 10 million years. Recent geochemical and seismic analysis of Uturuncu provides no evidence for a shallow magma chamber that would feed an imminent eruption, and the rising magma may never reach the surface. Nevertheless, many parties will certainly continue monitoring the volcano.

In addition to providing critical evidence to inform the magma-transport debate, the

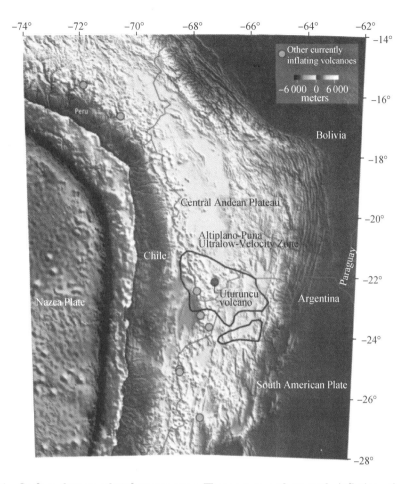

Fig. 7-1   Surface clues to subsurface processes. There are several currently inflating volcanoes in the Andes, including the Uturuncu volcano above the Altiplano-Puna Ultralow-Velocity Zone (APULVZ). On the basis of satellite data and computational modeling, Fialko and Pearse conclude that a magmatic diapir, sourced from the APULVZ, is rising below Uturuncu. It remains to be shown whether other inflating volcanoes are also fed by diapirs

occurrence of this diapir and its connection to the APULVZ places it in the middle of another fundamental discussion about how continental plateaus achieve their height. The diapir requires a large subsurface magma body to inflate the balloon continuously over decadal time scales. But how did the large magma body get there in the first place? A popular notion is that in the central Andes, crustal thickening due to the subduction of the Nazca Plate below the South American Plate caused the bottom of the continental crust to densify, "delaminate", and sink into the mantle. The isostatic adjustment from this rapid decrease in bulk density below the mountains caused the plateau to rise almost instantaneously by geologic standards. The thermal conditions set up by the purported delamination facilitated the APULVZ as a mid-crustal zone of partial melt. If the

29

delamination hypothesis is true，diapirs could be a necessary by-product.

The potential connection of the observed diapir to delamination，itself a result of a tectonic collision，reminds us that only a few decades have passed since the theory of plate tectonics was developed. Although Earth-orbiting satellites and computer simulations have replaced ship-towed magnetometers and graph paper as the tools of the moment，we are still sorting out the ramifications of modern Earth science's defining achievement.

（Cited from Brooks B A. Seeing is believing ［J］. Science，2012，338(6104)：207-208.）

New Words and Expressions

satellite	n.	卫星
computational	adj.	计算的
spheroidal	adj.	球状的
lava lamp	n.	熔岩灯
bureau	n.	办公桌
ponder	v.	仔细思考,沉思
density	n.	密度
granitic	adj.	花岗岩的
pluton	n.	深成岩体
magmatic	adj.	岩浆的
plausible	adj.	貌似合理的
interferometric	adj.	干涉测量的
aperture	n.	孔径、光圈
manifestation	n.	表现,显示
robust	adj.	鲁棒性的,稳定的
viscoelastic	n.	黏弹性
sophisticated	adj.	复杂的,精致的
diapir	n.	底辟
simultaneous	adj.	同时发生的
withdrawal	n.	收回,取消
viscosity	n.	黏性
silicic	adj.	含硅的
myriad	adj.	无数的,种种的
albeit	conj.	即使
prevalent	adj.	流行的,普遍的

ominous	*adj.*	不吉利的
eruption	*n.*	火山爆发
chamber	*n.*	洞室

译文:眼见为实

我儿子书桌上的熔岩灯给了他一个有利条件,促使他思考随着温度和密度的变化而升降的彩色斑点。地质学家可没有这样的特权。地球上一些最令人印象深刻的地质特征是岩浆运输的直接结果,例如加利福尼亚内华达山脉的巨大花岗岩岩体,但我们不能直接观察这些运输过程。在本期的第 250 页,Fialko 和 Pearse 使用卫星数据和计算模型来推断一个地点的传输过程。他们的结论是,一个岩浆底辟——一个大致球形的部分熔融的物质——正在从安第斯山脉中部的阿尔提普兰高原的部分区域升起。

为了推断地下演变过程,地质学家必须将地球表面每年的位移精确到毫米至厘米并与数学模型进行比较。然而,不同的模型可以在观测误差容许范围内给出相似的预测,并且物理上最合理的模型通常因为计算量太大,无法与数据进行充分的比较。因此,进一步的发展需要同时增强观测和计算能力。Fialko 和 Pearse 在他们出色的研究中做到了这两点。

作者使用合成孔径雷达干涉(InSAR)技术,结合来自三个不同卫星的数据,在玻利维亚乌图伦古火山建立了一个长达 18 年的地表位移记录。研究区位于目前被认为是地球大陆壳中最大的活动岩浆体的正上方——Altiplano-Puna 超低速带(APULVZ),深约 17~19 千米,厚约 1 千米,宽约 100 千米。

基于来自两个具有相似观测条件的卫星的 8 年 InSAR 数据,以往的研究识别出与乌图伦古火山有关的地表隆起,并通过建模将其作为深度 15~17 千米的膨胀岩浆点源的表现形式。为了更好地确定假定的岩浆源的深度、几何形状和时间历史,Fialko 和 Pearse 要求第三颗卫星从截然不同的方向获取数据。这个更有力的数据集与先前获得的数据一起,揭示了隆起周边地区出乎意料的沉降模式。

地球地壳弹性行为的假设很难解释组合的隆起和下沉。因此,Fialko 和 Pearse 测试了越来越复杂的黏弹性模型。他们令人信服的结论是,APULVZ 是向地表上升的底辟的岩浆源。上升的气球状底辟引起地表抬升,同时从 APULVZ 中抽出的物质引起周边沉降。

在熔岩灯中出现的底辟现象很容易被认同:我们可以看到它正在发生。但是,地壳是由黏度高得多的岩石组成的,关于底辟作用是否是一个可行的岩浆输送过程还没有达成共识。反对意见主要集中在缺乏由于底辟作用导致岩石强烈变形的现场证据。

这些反对意见引起了许多人的争论,他们认为硅质岩浆在叶状岩墙中迁移,这有大量的现场实例,尽管它们具有低黏度的铁和富镁的成分。另一方面,最近有更多热力学分析表明,我们不应放弃底辟的可能性,因为具有更多黏性的硅质成分的岩墙可能在距离其源区几百米内冻结。

Fialko 和 Pearse 的研究提供了最有力的证据来证明底辟的存在。显然,下一个问题是：它们有多普遍？InSAR 研究已经在世界各地发现许多与火山有关的膨胀信号，尤其是在安第斯山脉（图 7-1）。如果每一个中心都像乌图伦古火山那样彻底地研究，那么有多少会被称为"底辟"？然而，一个更关键的问题是，新发现会怎样影响灾害的发生？一个巨大而上升的岩浆体听起来不祥。乌图伦古火山已经约 270 000 年没有爆发，但在过去的 1 000 万年中该地区经历了大规模的喷发。根据最近乌图伦古火山的地球化学和地震分析结果，没有证据表明存在一个为即将发生的爆发提供岩浆来源的浅层岩浆室，且上升的岩浆可能永远不会到达地表。然而，许多团队肯定会继续监测火山。

除了为岩浆运输的争论提供关键的证据，这种底辟的出现及其与 APULVZ 的联系还使它处于另一个关于大陆高原如何达到其高度的基本讨论中。底辟需要一个大的地下岩浆体持续数十年使球体膨胀。但是巨大的岩浆体最初是如何到达那里的呢？一个普遍的观点是，在安第斯山脉中部，由于纳斯卡板块俯冲到南美洲板块之下，地壳增厚，导致大陆地壳底部致密，发生"分层"，并沉入地幔。根据地质学上的标准，由于山下部分容重迅速下降，均衡调整导致高原几乎瞬间上升。由上述的分层建立的热条件促使 APULVZ 变成中部地壳的部分熔融区域。如果分层假设是正确的，底辟可能是一个必要的"副产品"。

观测到的底辟与分层之间存在潜在联系，本身就是地壳构造碰撞的结果，这提醒我们板块构造理论提出以来仅仅过去了几十年。尽管地球轨道卫星和计算机模拟已经取代了船拖式磁强计和图纸成为当前的工具，我们仍然在整理现代地球科学成就的衍生成果。

Text 8　Taking Earth's Temperature

Much of what we know about temperature in the interior of Earth is inferred from phase transitions. Temperature within the rocky mantle（roughly between 30 km and 2,900 km depth）is of particular interest because it indicates the current state of Earth's heat engine，which powers nearly all geological processes. On Page 1,813 of this issue，van der Hilst and colleagues describe new results on the mantle temperature at two depths near the coremantle boundary（CMB）. The difference in temperature gives an estimate of the thermal gradient and heat flow at the base of the mantle. Such estimates are crucial for resolving long-standing questions about the distribution of heat-producing elements，the transport of heat through the interior，and the thermal evolution of the planet.

Changes in pressure and temperature with depth cause transitions in the crystal structure of mantle minerals，which can be detected as abrupt changes（or discontinuities）in the speed of seismic waves. Researchers typically infer mantle temperature from the depth of these discontinuities using an experimentally determined phase diagram. Discontinuities at depths of 410 km and 660 km fix the temperature in the upper mantle within experimental uncertainty, whereas the liquid-solid transition at the inner-core boundary anchors the temperature deep in the core. Our ability to interpolate temperature to other depths is compromised by large uncertainties in the core temperature and by an incomplete understanding of the radial structure of convection in the mantle.

To obtain their results, van der Hilst et al. used a seismic imaging technique that maps surfaces of abrupt change in wave speed. Seismic waves that reflect from surfaces near the base of the mantle are recorded mainly before the arrival of the principal reflection from the CMB (see the left panel of Fig. 8 – 1). To estimate the spatial structure, magnitude, and even the sign of wave-speed changes, the seismic data are processed by means of numerical methods called "inverse scattering". That is, instead of starting with a geophysical structure and calculating the scattering, the scattered waves are used to reconstruct the scattering structures. These methods were initially developed for hydrocarbon exploration, but growing interest in other areas of geophysics attests to the utility of high-resolution techniques when spatial sampling of the target region is sufficient.

The reflectors identified by van der Hilst et al. are located near the CMB below Central America. They attributed several prominent reflections to a newly discovered transition between the perovskite (pv) phase of the mineral (Mg, Fe) SiO_3, the most abundant mineral in the mantle, and the higher-pressure postperovskite (ppv) phase. Published experimental and theoretical estimates for the pressure ($P \approx 110$ to 125 GPa) and temperature ($T \approx 2,200$ to $2,700$ K) of the phase transition suggest that the ppv phase may be present near the base of the mantle. Moreover, the rate of change of pressure with temperature (the Clapeyron slope) is thought to be much larger than values typically reported for the other mantle transitions. A steep Clapeyron slope in the vicinity of a thermal boundary layer at the base of the mantle raises the possibility of a double crossing of the pv-ppv phase boundary. This first transition occurs at the depth where pressure converts pv into the more compact ppv structure. The resulting density increase of 1% to 2% is accompanied by a 2% to 4% increase in the speed of shear waves. A second transition occurs within the boundary layer, where a rapid increase in temperature converts the ppv phase back to pv (see the right panel of Fig. 8-1).

Reflection of shear waves from a double crossing of the pv-ppv phase boundaries should produce characteristic reflections, and the observations support this expectation. The data also reveal substantial variations in the height of the reflectors above the CMB. The upper transition appears to be deflected upward in a region where subduction of the Cocos plate of Central America would presumably lower the temperature. The observed topography is consistent with the positive Clapeyron slope of the pv-ppv transition. The

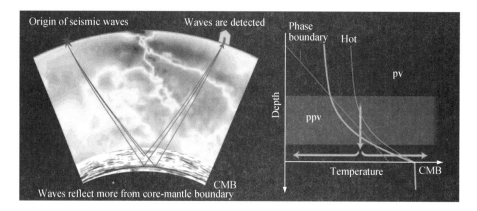

Fig. 8-1　Core values. (Left) Waves from a seismic disturbance reflect from different structures in the CMB and are detected at the surface. (Right) Intersection of temperature curve (thick yellow line) with the pv phase transition boundary (green line) defines the upper and lower surfaces of the ppv layer. No transition occurs when the overlying mantle is sufficiently hot (thin yellow line). The dense ppv layer (shaded) drives flow toward and along the CMB, altering the temperature gradient due to the effects of latent heat

lower transition appears to be deflected downward in this cold region, which is reasonable given that temperature in the boundary layer must adjust to a nearly constant value at the surface of the core; a steeper thermal gradient crosses the phase transition closer to the CMB.

At the margins of the study area, the upper and lower transitions appear to merge, implying that these regions are too hot to permit the ppv phase. A similar structure (e.g., a region of ppv phase) is present in the central Pacific, suggesting that double crossings are widespread but not ubiquitous.

So what does this finding mean for Earth's thermal state? The occurrence of a double crossing anywhere indicates that the temperature at the top of the core must exceed the transition temperature at the pressure of the CMB. A lower core temperature permits only a single reflector that must be present everywhere above the CMB.

Conversely, a higher core temperature permits intermittent regions of ppv phase in places where the radial geothermal gradient (rate of change of temperature versus radius) is steeper than the phase boundary (-5 K/km to -8 K/km). Regions where the ppv phase is absent indicate a smaller radial gradient and less heat flow. More specific estimates of temperature and heat flow can be obtained by fitting a simple parametric representation of the boundary layer temperature. By matching the points of intersection of thermal gradient with the phase boundary to the height of the observed reflectors, it is possible to calculate the temperature and heat flow at the CMB. Of course, the phase boundary is not well known, and the influence of compositional variations is currently a wild card, but the general relationship between the phase boundary and the local temperature profile should stand the test of time, as long as the reflectors are correctly interpreted as phase transitions. As our knowledge of the phase diagram improves, so will our knowledge of temperature near the base of the mantle.

Another wrinkle is introduced when the dense layer of ppv sinks into the less dense layer of pv. A settling velocity of 1 mm/year is plausible given typical (but uncertain) estimates for the density contrast, viscosity, and characteristic dimensions of the ppv region, including its height above the CMB. Heat is absorbed at the base of the zone by conversion of ppv to pv, steepening the temperature gradient at the CMB, possibly by a factor of two or more relative to the predictions obtained with the simple boundary-layer model. A higher heat flow would cool the underlying core more rapidly and drive more vigorous convection, increasing the power available to generate Earth's magnetic field. However, additional energy sources in the core may be needed to maintain the higher heat flow.

The important result in this study is the detection of new phase transitions in the mantle with seismic methods that have only recently been applied to the study of the deep interior. Advances in the methodology are likely and better data coverage is expected through the ongoing USArray component of Earthscope. Future work with this and other instruments has the potential to reveal new insights into the inner workings of our planet.

(Cited from Buffett B A. Taking earth's temperature [J]. Science, 2007, 315(5820): 1801—1802.)

New Words and Expressions

interior	*adj.*	内部的
mantle	*n.*	覆盖层,地幔
thermal	*adj.*	热的
gradient	*n.*	梯度
seismic	*adj.*	地震引起的,与地震有关的
interpolate	*v.*	内插
convection	*n.*	对流
obtain	*v.*	获得
spatial	*adj.*	空间的
magnitude	*n.*	量,大小,幅度
scatter	*v.*	分散
hydrocarbon	*n.*	烃,碳氢化合物
attest	*v.*	表明,认证
prominent	*adj.*	突出的

perovskite	*n*.	钙钛矿
topography	*n*.	地形,地貌
margin	*n*.	余量,余额,页边
ubiquitous	*adj*.	普遍的
intermittent	*adj*.	断续的
plausible	*adj*.	可能的
viscosity	*n*.	黏质,黏性
vigorous	*adj*.	旺盛的,强烈的

译文:测量地球的温度

我们对地球内部温度的大部分了解都是从相变过程推断出来的。岩石地幔(深度约为30千米至2 900千米的部分)内的温度特别令人感兴趣,因为它表明了地球热源的当前状态,而热源为几乎所有的地质过程提供动力。在本期第1 813页,van der Hilst及其同事描述了地核—地幔边界附近(CMB)两个深度的地幔温度的新结果。通过温差可以估算出地幔底部的热梯度和热流,而这些估算结果对于解决产热元素分布、内部热量传输以及行星热演化等长期问题至关重要。

压力和温度随着深度变化而变化导致了地幔矿物结晶体结构的转变,这种变化可从探测到的地震波速度的突然(或不连续的)变化体现出来。研究人员通常使用实验确定的相图从深度不连续处推断地幔温度。在410千米和660千米深度处的不连续性确定了实验未能确定的上地幔的温度,而内核边界处的固液相变确定了核内部的温度。由于核温度存在较大的不确定性和对地幔径向对流了解得不够完全等原因,我们通过内插法得到其他深度温度的能力受到了限制。

为了获得结果,van der Hilst等人利用地震成像技术绘制了波速突然变化的表面。从地幔基部附近的表面反射的地震波主要记录在来自CMB的主反射到达之前(图8-1左侧)。为了估计波速变化的空间结构、幅度、甚至信号,将地震数据通过一种叫做"逆散射"的数值方法进行处理。此方法并不用地球物理结构来计算散射波,而是由散射波数据重建散射结构。这些方法原本用于油气勘探,但是在其他地球物理学领域日益增长的兴趣证明了当目标区域空间采样足够时高分辨率技术的实用性。

van der Hilst等人确定的反射面位于中美洲以下的CMB附近。他们将几种突出的反射信号归因于$(Mg, Fe)SiO_3$(地幔中最丰富的矿物)的钙钛矿(pv)相和高压后钙钛矿(ppv)相之间新发现的转变过程。已发表的对相变所需的压力($P \approx 110 \sim 125$ GPa)和温度($T \approx 2\,200 \sim 2\,700$ K)所做的实验和理论估算表明,ppv相可能存在于地幔基部附近。此外,压力随温度的变化率(Clapeyron斜率)被认为远大于通常报告的其他地幔转变的值。在地幔基部的热边界层附近陡峭的Clapeyron斜率提高了pv-ppv相界双重交叉的可能性。第一次相变发生在压力将pv转化为更紧凑的ppv结构的深度处,由此产生的密度增加1%至2%

并伴随着剪切波速度增加 2% 至 4%；第二次相变发生在边界层内，其中温度的快速升高将 ppv 相转变回 pv（图 8-1 右侧）。

pv-ppv 相界的双重交叉所反射的剪切波应当会产生特征反射，而观察结果也支持这种预测。数据还证明了 CMB 上方反射面高度的显著变化。在可能由于中美洲 Cocos 板块的俯冲而降温的区域，上部的转换似乎会向上偏转。观察到的地形与 pv-ppv 跃迁的正 Clareyron 斜率一致，下部的转换似乎在这个寒冷区域向下偏转；这种现象是合理的，因为边界层中的温度必须在核心表面调整到几乎恒定的值，在靠近 CMB 的相变过程中热力梯度的变化是很大的。

在研究区域的边缘，上部和下部的转换似乎是合并在一起的。这意味着温度高到 ppv 相已无法存在。类似的结构（例如 ppv 相的区域）存在于中太平洋，表明双交叉是广泛的但不是普遍存在的。

那么这个发现对于地球的温度状态意味着什么呢？双交叉的出现区域证明，核心顶部的温度必须超过 CMB 压力下的相变温度。较低的核心温度只允许单个反射面存在于 CMB 上方的某个位置。

相反，较高的核心温度则允许不连续的 ppv 相区域存在于径向地热梯度（温度与半径的变化率）比相界（-5 K/km 至 -8 K/km）更陡的地方；而没有 ppv 相的区域意味着较小的径向梯度和热流。通过拟合边界层温度的简单参数表示，可以获得更具体的温度和热流估计。通过匹配相界处的热力梯度与观察到的反射面的高度，可以计算 CMB 处的温度和热流。当然，相界不是完全清楚的，而且成分变化带来的影响也难以预料。但是，只要将反射面正确地理解为相变，相界和局部温度曲线之间的一般关系就能经得起时间的考验。随着对相图相关知识的增加，我们对地幔基部的温度的认识也会加深。

当致密的 ppv 层沉入不那么致密的 pv 层时，会产生另一种揉皱。如果给定 ppv 区域的密度对比、粘度和特征尺寸的典型（但不确定）估计，包括其在 CMB 上方的高度，1 毫米/年的沉降速度是合理的。ppv 转换为 pv 时会在区域的基部吸收热量，使 CMB 处的温度梯度变陡，对比由简单边界层模型获得的预测可能会差上两倍甚至更多。较高的热流会更快地冷却下面的核心，并驱动更强烈的对流，增加产生地球磁场的能量。然而，较高的热流也可能需要核心中的额外能量来维持。

这项研究的重要成果是利用最近才应用于深部研究的地震法来探测地幔中的新相变。该方法仍很可能有新进展，并且正在发展的美国地震台阵地球透镜组件也预计会扩大数据的覆盖范围。未来使用这些以及其他仪器的工作有可能对地球内部的运作规律提出新的见解。

Text 9　Magnetostatics

Magnetic field originates when a charge moves. Therefore no magnetic field is associated with electrostatics. Magnetic field has link with direct and alternating current flow fields. Magnetism, (i.e., the property of certain metallic objects, to attract or repel some other metallic objects) was known to the people for the last several hundred years.

The word "magnetism" came from the word "Magnesia", an ancient city of Asia Minor. Certain rocks in the vicinity of this city had the property of attracting metallic bodies. It was observed that a needle shaped load stone got deflected along a particular direction irrespective of any arbitrary orientation and it was used by mariners to find out the north-south direction. This kind of movement in the needle is possible when a couple act on it. The presence of a couple is possible, when a field exists in the north-south direction and the needle has north and south polarity at the two ends. Thus the existence of the geomagnetic field was conceptualised.

In 1819 Oersted first observed that there is a close connection between the flow of electric current through a wire and the generation of magnetic field. In 1820 Biot and Savart first experimentally demonstrated the quantitative aspect of strength of the magnetic induction B and magnetic field H. In the same year Ampere proposed his force law i.e., the law for force between the two coils carrying currents.

Magnetic field is a global and naturally occurring field like gravity field and can be measured anywhere on the surface of the earth, in the air, in the ocean bottom and inside a borehole. Both magnetic and gravity fields show local perturbations due to local variations in magnetic susceptibility and density. Geophysicists are interested about these local and global perturbations.

Gravity field generates always a force of attraction but magnetic field, like electrostatic field, can have either the force of attraction or repulsion according to the law "like poles repel and unlike poles attract". Magnetic field has conceptual north and south poles, the way we have positive charge and negative charge in electrostatics. In this particular aspect magnetostatic field has some similarity with the electrostatics field i.e., both the fields satisfy Coulombs law. Electrostatic field, magnetostatic field and gravity field follow inverse square law. The fields vary directly with the product of charges or masses or pole strengths and inversely as the square of the distance. The constants of proportionality are different for different fields. Both in the case of electrostatic field and direct current flow field, we brought the concept of potential and electromotive force using the line integral of force multiplied by distance. Similar concept of magnetomotive force exists where the work is done in the magnetic field and the line integral of the magnetic field times the distance gives magnetomotive force. Magnetostatics has the concept of both scalar and vector potentials as well as rotational and irrotational field. Irrotational nature of the magnetic field comes from the low frequency approximation and in a source free region.

Positive charge and negative charge in the case of electrostatics, source and sink in the case of direct current flow field can generate both bipole and dipole fields depending upon the separation of the two opposite charges or two opposite current sources. Separation between the north pole and south pole can generate bipole and dipole fields in magnetostatics. That way magnetostatic field has some similarities with the electrostatic field and direct current flow field.

Magnetostatic field has significant dissimilarities with the other fields. Magnetostatic field is always a bipole or a dipole field. An isolated north pole or south pole does not exist. A coil carrying current or a thin sheet of magnetic substances with negligible thickness have both north and south poles.

Magnetostatic field is a solenoidal field. Divergence of the magnetic field is always zero because the poles do not stay in isolation. In the case of electrostatic field, direct current flow field, gravity field, heat flow field etc, if the area under consideration is a source free region, then they become solenoidal field and satisfy Laplace equation. If the region under consideration contains source, then the fields will be nonsolenoidal. These fields then satisfy Poisson's equation. So both the options are there in the said nonmagnetic fields.

Magnetostatic field is a rotational field (Figs. 9-1, 9-2). The curl of a magnetic field is not zero where as curl of gravity, electrostatic, direct current flow, heat flow fields etc. are zero. These fields are irrotational field. Time varying electromagnetic field also is a rotational field. In the absence of any current source magnetostatic field can also be an irrotational field. Geomagnetic field, because of its low frequency approximation, becomes an irrotational field and satisfy Laplace equation.

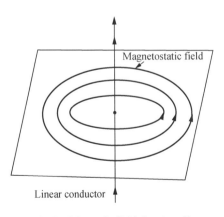

Fig. 9-1   **Magnetic field due to a linear conductor carrying current**

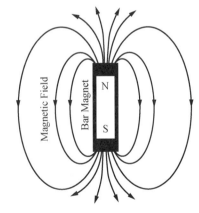

Fig. 9-2   **Magnetic field due to a bar magnet**

Amongst all the static and stationary fields only magnetostatics has the concept of vector potential. This is also one of the major differences with other fields. These vector potentials received greater attention while solving the electromagnetic boundary value problems in geophysics.

Magnetic lines of forces are continuous where as the electric field in electrostatics starts from a positive charge and ends in a negative charge; current flow lines in direct current flow field starts from a source and ends in a sink. In this respect also magnetic field has dissimilarities with the electrostatic field and the direct current flow field. The concept of bipole and dipole exists in magnetostatics, electrostatics and DC field. A coil carrying current is called a magnetic dipole. A coil carrying alternating current is termed as an oscillating magnetic dipole. Magnetostatic field, electromagnetic field, gravity field can be measured in the air. Aeromagnetic, aeroelectromagnetic and aerogravity methods are standard geophysical airborne tools. For direct current flow however galvanic contact of both current and potential measuring probes with the ground or any other medium of finite conductivity is necessary. Therefore DC flow field does not have any airborne counterpart.

(Cited from Roy K K. Potential theory in applied geophysics [M]. 1st ed. Berlin Heidelberg: Springer Science & Business Media, 2008.)

New Words and Expressions

magnetostatics	*n.*	静磁学
charge	*n.*	电荷
couple	*n.*	一双,成对;力偶
arbitrary	*adj.*	随意的,任意的
orientation	*n.*	方向
conceptualize	*v.*	概念化
direct current flow	*n.*	直流
perturbation	*n.*	扰动,干扰
solenoidal field	*n.*	无散度场
source	*n.*	发射源,源头
oscillating	*adj.*	摆动的,振荡的
galvanic	*adj.*	电流的

译文:静磁学

当电荷移动时会产生磁场,因此磁场与静电场无关,而与直流和交流电场有联系。人们在几百年前已经意识到了磁性(即某些金属物体吸引或排斥其他金属物体的性质)的存在。

"磁"这个词来源于古老的小亚细亚城市 Magnesia。这个城市附近的某些岩石具有吸引金属物体的特性。人们发现,无论将针状磁石的朝向设置为什么方向,它都会向特定的方向偏转。水手们利用这一性质来辨别南北方向。磁针移动是因为受到了力偶的作用。这个力偶的来源是因为磁针位于一个具有南北方向的磁场中,而且磁针的两端具有南北极性。由此,人们确立了地磁场的概念。

1819 年,Oersted 首次发现,通过导线的电流能够产生磁场。在 1820 年,Biot 和 Savart 首次通过实验证明了磁感应强度 B 和磁场 H 之间的定量关系。同年,Ampere 提出了两个承载电流的线圈之间的作用力定律。

像重力场一样,磁场是一个在地球上自然存在的场。它可以在地球表面、空气中、海底和孔洞内的任何地方被检测到。受到局部磁化率和密度变化的影响,磁场和重力场都存在局部异常现象。地球物理学家对这些局部和全球性的异常很感兴趣。

重力场只能产生引力,但磁场类似于静电场,可以产生引力和斥力,并且遵循"同极相斥、异极相吸"的规律。正如静电场中正负电荷的定义方式,人们在磁场概念里确定了南北极。在这方面,静磁场与静电场具有一些相似性,比如两者都满足库仑定律。静电场、静磁场和重力场都遵循平方反比定律,即这些场的强度直接正比于电荷(或磁感应强度、质量),而反比于距离的平方。不同场的比例常数是不同的。对于静电场(直流电流场),我们使用力的线积分与距离的乘积来确定势(电动势)的概念。在磁场中通过类似的工作,我们得到了磁动势的概念,即磁场的线积分乘以距离。磁静电学既有标量势和矢量势的概念,也有有旋场和无旋场的概念。磁场的无旋性质来源于低频近似和无源区域。

根据性质相反的两个电荷或流向相反的两个电流源的分离情况,静电场中的正负电荷以及直流电流场中的流出源和流入点可以产生双极场和偶极场。而南北极的分离可以在静磁场中产生双极场和偶极场,说明静磁场与静电场和直流电流场之间具有某些相似之处。

静磁场与其他场的一个显著的区别在于,静磁场总是双极场或偶极场。不存在单独的北极或南极。即使是承载电流的线圈或厚度可以忽略不计的薄磁性物质也都具有北极和南极。

静磁场是无散度场。由于没有孤立的磁极,磁场的散度总是为零。在静电场、直流电场、重力场、热流场等情况下,如果所考虑的区域是无源的,则它们成为无散度场且满足拉普拉斯方程;如果所考虑的区域是有源的,那么这些场将是有散度场,并且满足泊松方程。因此,在上述的非磁场中存在两种可能的情况。

静磁场是有旋场(图 9-1,图 9-2)。磁场的旋度不为零;而重力、静电场、直流电流场、热流场等场的旋度为零,因此这些场是无旋场。随时间变化的电磁场也是有旋场。如果没有任何电流源,静磁场也会变成无旋场。地磁场,由于其低频近似性,可以视为无旋场并满足拉普拉斯方程。

在所有静态和静止场中,仅静磁场具有矢量势的概念。这也是它与其他场的主要差别之一。矢量势的这一性质在解决地球物理学中的电磁边界值问题时得到了更多的关注。

磁力线是连续的,而静电场中的电流线从正电荷开始,至负电荷结束;直流电流场中的电流线从流出源开始,至流入点结束。在这一点上,磁场不同于静电场和直流电场。在静磁场、静电场和直流电流场中均有双极和偶极子的概念。承载电流的线圈被称为磁偶极子,承载交流电的线圈被称为振荡磁偶极子。人们可以通过航磁、航空电磁和航空重力方法等标准的空中地球物理探测手段从空中测量静磁场、电磁场、重力场。然而,要想测量直流电流场,电流和电位的测量探针必须与地面或其他具有一定导电率的介质有电流的接触,因此人们无法从空中测量直流电流场。

Text 10 Geophysical Investigation Methods

Borehole georadar experiments

Borehole georadar data were recorded in each of the three deep boreholes using both single-hole radar reflection and vertical radar profiling (VRP) methods and crosshole techniques. These borehole methods allowed the electromagnetic reflection characteristics of the rock mass to be investigated at depths greater than could be reached using surface georadar approaches. Borehole reflection methods are particularly useful for imaging steeply dipping fractures as shown in Fig. 10-1(a) (Olsson et al., 1992; Tronicke and Knoll, 2005). At Randa, it involved the progressive movement of a fixed-offset transmitter-receiver antenna pair (MALA system with 100 MHz antennas) along the borehole. The interpretation of the reflections that intersected the borehole walls was constrained by the televiewer images (Spillmann et al., 2007). Fortunately, some reflections that did not cut the boreholes and hence could not be directly oriented (single-hole georadar images by themselves contain no azimuthal information as to the direction

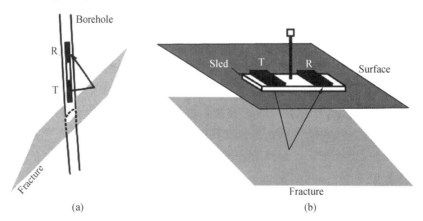

(a)   (b)

Fig. 10-1 Illustration of borehole and surface georadar methods employed: (a) Single-hole georadar reflection measurement. The radar system, which contains both a transmitter (T) and receiver (R) antenna, is moved along the borehole axis taking measurements every 10 or 20 cm (modified after Spillmann et al., 2007a); (b) Surface georadar surveys in which the transmitter and receiver antenna are mounted on a sled together with a prism for determining the 3-D position of the sled at each measurement point using a self-tracking theodolite (modified after Heincke et al., 2005)

of the reflector) could be projected to faults and fracture zones mapped on surface, thus allowing their orientation to be derived.

The VRP data were gathered in SB 50S and SB 50N along a surface profile between the boreholes, and in SB 120 to a depth of 50.5 m using a surface line that extended out to 28.5 m from the hole towards the north (see Fig. 10-2). Georadar crosshole tomography between the two 50 m deep boreholes yielded estimates of radar velocity and attenuation between the boreholes (Spillmann et al., 2007). The results of the VRP and crosshole experiments are not reported in this paper as they provided only limited useful data (Spillmann et al., 2007).

3-D surface georadar

The 3-D surface georadar technique has the potential to image faults and fracture zones within crystalline rock masses with high resolution and accuracy (Grasmück, 1996), but the penetration depth is usually limited to several tens of meters due to attenuation and scattering effects. The Randa surface georadar data were acquired across two overlapping 480—850 m^2 areas (Fig. 10-2). Two unshielded 100 MHz antennae mounted on a sled [Fig. 10-1(b)] were used to record data along densely spaced parallel lines (Heincke et al., 2005). After pre-processing, migration of the data accounting for topography produced images of moderately to shallowly dipping features (Heincke et al., 2005). To map steeply dipping structures characterised by rough surfaces, a new scheme that highlighted the presence of diffractions was developed and applied to the georadar data (Heincke et al., 2006a). This new semblance-based migration scheme, which also accounts for the influence of topography, was effective in mapping the locations of numerous steeply dipping structures throughout the depth range of interest.

3-D seismic refraction tomography

Unlike the georadar techniques, the 3-D tomographic seismic refraction method does not have the resolution to determine the locations and orientations of discrete fault and fracture zones. However, the distribution of low and high P-wave velocities within the rock mass is a measure of the degree of fracturing. Several case studies using 2-D refraction tomography on unstable rock slopes have previously been reported (Hack, 2000; Godio et al., 2006). To our knowledge, our survey was the first to involve a true 3-D approach (Heincke et al., 2006b).

The Randa seismic survey was designed to cover accessible parts of the unstable rock mass and large regions of the presumed stable mountain slope (Fig.10-2). The primary layout involved eight profiles, five oriented in an E-W direction and 3 oriented in a N-S direction. Source and receiver spacings along the 126-324 m long profiles were 4 and

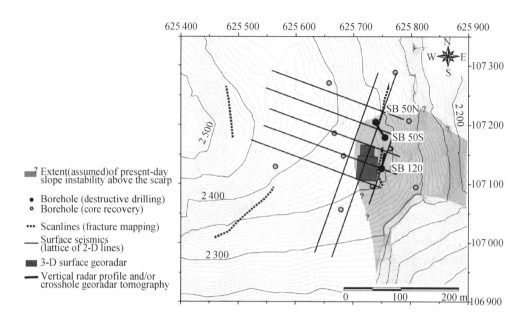

Fig. 10-2 Location of the geological and geophysical survey areas. The focus of the investigations was the upper part of the unstable rock mass, where the three deep exploratory boreholes (SB 120, SB 50S and SB 50N) were drilled

2 m, respectively. By using seven 24-channel 24-bit Geode recording systems, we were able to deploy receivers along the entire length of each profile (i.e. 64-163 receivers depending on the length of profile). To minimise disturbance to the surrounding alpine environment, small shot charges of 5-50 g were detonated in shallow (0.5-0.7 m deep) holes. In addition to the shotholes along the profiles, we also drilled clusters of 8 holes (1 per profile) at 33 locations offset from the profiles. Signals generated by shots detonated at these clusters were recorded on receivers along all profiles (Heincke et al., 2006b). Furthermore, all inline and offset shots were recorded by 3-component geophones deployed in the three moderately deep boreholes and in the nine shallow boreholes from which core samples were obtained (locations shown in Fig. 10-2). Together, the profiles and offset shots and receivers yielded moderately good to very good coverage of the entire investigation volume (Heincke et al., 2006b).

The P-wave first arrival times were inverted using a 3-D tomography algorithm. Comparison of estimated seismic velocities with those measured on intact core samples allowed Heincke et al. (2006b) to estimate the Seismic Rock Quality Designation (SQRD) (Hack, 2000) of the rock mass.

(Cited from Willenberg H, Loew S, Eberhardt E, et al. Internal structure and deformation of an unstable crystalline rock mass above Randa (Switzerland): Part I-Internal structure from integrated geological and geophysical investigations [J]. Engineering Geology, 2008, 101(1-2): 1-14.)

New Words and Expressions

fractures	*n.*	断裂,破裂	
azimuthal	*adj.*	方位角的	
orientation	*n.*	方向	
tomography	*n.*	断层摄影术	
crystalline	*adj.*	晶体的,结晶的	
penetration	*n.*	渗透	
attenuation	*n.*	衰减	
sled	*n.*	滑坡	
migration	*n.*	迁移	
shotholes	*n.*	爆破孔	
detonate	*n.*	爆炸	

译文:地球物理勘探方法

钻孔地质雷达试验

利用单孔雷达反射和垂直雷达剖面(VRP)法以及井间技术,技术人员记录了三个深井中每个井的钻孔地质雷达数据。由于地表地质雷达法的探测深度有限,这些钻孔探测方法使得研究深部岩体的电磁反射特性成为可能。钻孔反射法对陡峭裂缝的成像特别有用,如图 10-1a 所示(Olsson et al.,1992;Tronicke and Knoll,2005)。在 Randa,这种方法采用的是将一对固定偏移的发射—接收天线(具有 100 MHz 天线的 MALA 系统)沿钻孔逐步移动的方式。对穿越井壁的反射信号的解译受到观测图像的限制(Spillmann et al.,2007)。幸运的是,一些没有穿过钻孔而不能直接定向的反射信号(单孔地质雷达成像本身不包含关于反射器方向的方位角信息)可以投射到映射在地表的断层和断裂带上,进而使人们求得它们的方向。

VRP 数据是在 SB 50S 和 SB 50N 钻孔之间沿着一个表面轮廓获得的,我们还在 SB 120 钻孔处利用一条从钻孔位置向北延伸 28.5 m 的表面线收集了深度达到地下 50.5 m 处的数据(图 10-2)。我们利用两个 50 米深的钻孔之间的地质雷达井间层析成像数据,得到了钻孔之间雷达速度和衰减的估计值(Spillmann et al.,2007)。本文未报导 VRP 和井间试验的结果,因为它们只能提供有限的有用数据(Spillmann et al.,2007)。

3D 地表地质雷达

3D 地表地质雷达技术在结晶岩体内的断层和断裂带的高分辨率及高精度成像方面具有优势(Grasmück,1996),但受限于衰减和散射效应,它的信号穿透深度往往只有几十米。

Randa 的地表地质雷达数据是在两个重叠的 480~850 平方米的区域内获得的(图 10-2)。技术人员利用安装在滑板上的两个开放的 100 MHz 天线(图 10-1b),沿着密集排列的平行线记录数据(Heincke et al., 2005)。在预处理之后,我们可以通过对地形数据的迁移生成具有中度到浅度倾斜特征的图像(Heincke et al., 2005)。为了绘制具有粗糙表面特征的陡峭的倾斜结构,人们提出了一种突出衍射现象存在的新方案,并将其应用于地质雷达数据(Heincke et al., 2006a)。这种基于表象的新迁移方案也考虑了地形的影响,能有效绘制出探测深度范围内的大量的陡峭倾斜结构的位置。

3D 地震折射层析成像

与地质雷达技术不同,3D 地震折射层析成像技术没有足够的分辨率来确定离散断层和断裂带的位置和方向。然而,岩体内不同位置处的 P 波的波速差异可以用来评估其碎裂程度。已有文献报导了利用 2D 折射层析成像技术研究不稳定岩石边坡的若干个案例(Hack,2000;Godio et al., 2006)。据我们所知,我们的勘测是第一次应用了真正的三维技术(Heincke et al., 2006 b)。

Randa 的地震勘测范围覆盖了不稳定岩体的可勘测部分和被认为稳定山坡的大部分区域(图 10-2)。主要的布局分为八个剖面,其中五个为东西朝向,三个为南北朝向。沿着 126~324 米长的剖面,发射源和接收器的间距分别为 4 米和 2 米。通过使用 7 个 24 通道的 24 位 Geode 数据记录系统,我们能够在每个剖面的整个长度上布置接收器(即根据剖面的长度布置 64~163 个接收器)。为了将对周围高山环境的干扰最小化,我们在浅孔(0.5~0.7 米深)中引爆了少量(5~50 g)的炸药。除了沿剖面的孔外,我们还在偏离剖面的 33 个位置上设置了 8 个钻孔(每个剖面 1 个)。沿着剖面布置的接收器可以记录在这些钻孔中爆破产生的信号。此外,在 3 个中等深度和 9 个浅部取样孔中布置的三分量地震检波器(位置如图 10-2所示)记录了所有在线内的和偏移的爆破信号。总体而言,剖面、偏移爆破和接收器对整个勘察区域的覆盖效果相当不错(Heincke et al., 2006b)。

P 波的初次到达时间可以根据 3D 成像算法反推得到。Heincke 等人(2006b)通过比较预测的弹性波速和在完整岩芯样本中测定的弹性波速来推测岩体的地震岩石质量指标(SRQD)(Hack,2000)。

Unit 3 Hydrology

Text 11   A Decade of Sea Level Rise Slowed by Climate-Driven Hydrology

---

Over the past century, sea level rose at an average rate of 1.5 ± 0.2 mm year^{-1}, increasing to 3.2 ± 0.4 mm year^{-1} during the past two decades. The increase in the rate of rise is attributed to an increase in mass loss from glaciers and ice sheets and to ocean warming. Although these contributions are fairly well constrained, trends in sea level also contain a land water storage component that is acknowledged to be among the most important yet most uncertain contributions, in which land water storage is defined by the Intergovernmental Panel on Climate Change (IPCC) as all snow, surface water, soil moisture, and groundwater storage, excluding glaciers. Every year, land temporarily stores then releases a net $6\,000 \pm 1\,400$ Gt of mass through the seasonal cycling of water, which is equivalent to an oscillation in sea level of 17 ± 4 mm. Thus, natural changes in interannual to decadal cycling and storage of water from oceans to land and back can have a large effect on the rate of sea level rise (SLR) on decadal intervals. From 2003 to 2011, SLR slowed to a rate of ~ 2.4 mm year^{-1} during a period of increased mass loss from glaciers and ice sheets. Climate-driven changes in land water storage have been suggested to have contributed to this slowdown, but this assertion has not been verified with direct observations.

Until recently, little data have existed to constrain land water storage contributions to global mean SLR. As a result, this term has either been excluded from SLR budgets or has been approximated by using ad hoc accounting that includes modeling or scaling of a variety of groundbased observations. Human-induced changes in land water storage (hereafter referred to as "human-driven land water storage") include the direct effects of groundwater extraction, irrigation, impoundment in reservoirs, wetland drainage, and deforestation. These activities may play a major role in modulating rates of sea level change, and several studies of large aquifers suggest that trends in regional and global land water storage are now strongly influenced by the effects of groundwater withdrawal. Currently, human activity (including groundwater depletion and reservoir impoundment) is estimated to have directly resulted in a net 0.38 ± 0.12 mm year^{-1} sea level equivalent (SLE) between 1993 and 2010 or 15 to 25% of observed barystatic SLR, but estimates are acknowledged to have large uncertainties.

Climate-driven variability in rainfall, evaporation, and runoff also contributes to decadal rates of sea level change through changes in the total amount of water held in snow, soil, surface waters, and aquifers. Climate-driven changes in land water storage have been assumed to be too small to include in sea level budgets, but there is little observational evidence to support this assumption. The vast spatial scale of climate-driven changes in land water storage has made them too difficult to observe with accuracy. As such, current IPCC sea level budgets exclude a potentially large water storage term that is required for closure of barystatic SLR on decadal time scales.

We assessed the role of land water storage in SLR over the 12-year period from 2002 to 2014. We examined global changes in surface mass derived from satellite measurements of time-variable gravity that are well-suited to constrain global changes in water storage. From this data, we extracted an observation-based estimate of the net contributions of the continents to SLR. By incorporating recently reconciled estimates of glacier losses [an update to Gardner et al.] and recent estimates of global groundwater depletion, we are able to disaggregate this net mass change into the contributions of glaciers, direct human-driven land water storage, and climate-driven land water storage.

Measurements of time-variable gravity come from NASA's Gravity Recovery and Climate Experiment (GRACE) satellite mission. GRACE provides monthly observations of changes in the Earth's gravity field that, after the removal of signals owing to changes in solid earth and atmosphere, result from the movement of water and ice through the Earth system at specific temporal and spatial scales. GRACE has provided monthly gravity field solutions since April 2002 and has proved to be an effective tool with which to observe changes in the mass of ice sheets, glaciers, snow mass, regional groundwater storage, and surface water storage. Previous studies have shown that because the accuracy of GRACE measurements generally increases with the size of the domain, GRACE observations may be useful to constrain hydrology contributions to sea level change, although only Rietbroek et al. have attempted to disaggregate those contributions by process. The increasing length of the GRACE record, combined with recent improvements in the processing of the intersatellite range-rate measurements and modeling of gravity change resulting from changes in solid earth displacement (supplementary materials, materials and methods), have now made the GRACE record more relevant to investigation of land water storage contributions to sea level.

(Cited from Reager J T, Gardner A S, Famiglietti J S, et al. A decade of sea level rise slowed by climate-driven hydrology [J]. Science, 2016, 351(6274): 699-703.)

New Words and Expressions

reconcile	v.	调解
disaggregate	n.	分解,分散
gigaton	n.	十亿吨
oscillation	n.	波动,振荡
ad hoc	adj.	特别的,临时的
aquifer	n.	含水层
depletion	n.	消耗
impoundment	n.	蓄水

译文:气候引起的水文变化减缓了十年间海平面的上升速度

在过去一个世纪里,海平面以1.5±0.2毫米/年的平均速度在上升;而在最近二十年内,海平面的上升速度已升至3.2±0.4毫米/年。上升速度的提高是冰川、冰盖的融化和海洋变暖引起的。虽然这些因素的作用非常明确,但是海平面的变化趋势也包含一个被公认是最重要但最不确定的因素:陆地水储存量。根据政府间气候变化专门委员会的定义,陆地水储存量包括除冰川外的所有的雪、地表水、土壤水和地下水。每年,陆地上临时储存的水通过季节性的水循环释放出净60 000±14 000亿吨的质量,相当于海平面17±4毫米的波动。因此,从年际到年代际的海洋和陆地间往复循环的水储存量的自然变化,会对十年间海平面的上升速率产生很大的影响。从2003年到2011年,冰川和冰盖的融化质量在增加,但海平面的上升速率反而降至2.4毫米/年。气候因素引起的陆地水储存量的变化被认为是造成海平面上升减缓的原因,但是这种说法并没有被直接观测到的结果证实。

至今还几乎没有数据能够证明陆地水储存量的变化会影响全球平均海平面的上升。因此,这个因素要么被排除在海平面上升的收支因素之外,要么需要使用特定的方法来进行估算,包括对各种大地观测数据进行建模或扩展。人类活动引起的陆地水储存量的变化(以下简称"人类引起的陆地水储存")包括地下水开采、灌溉、水库蓄水、湿地排水和森林砍伐等活动的直接影响。这些活动在调节海平面变化速率方面可以发挥主要作用。对大型含水层的若干研究表明,区域和全球范围内的陆地水储存量的变化趋势正在遭受地下水开采的强烈影响。据估计,目前人类活动(包括地下水消耗和水库蓄水)直接导致海平面在 1993 年至 2010 年间产生了净 0.38±0.12 毫米/年的上升速率,占观测到的海平面上升速率的 15% 至 25%,但是此估计值具有很大的不确定性。

受气候的影响,降雨量、蒸发量和径流量都会产生波动,并引起在积雪、土壤、地表水和地下含水层中蓄存的总水量发生变化,进而影响十年间海平面的变化速率。受气候影响的陆地水储存量的变化被假设为很小而没有纳入海平面变化的收支因素中,但是几乎没有观测证据能够支持这一假设。由于气候影响导致的陆地水储量的变化在空间尺度上分布非常广泛,它们很难被准确地观测。因此,目前政府间气候变化专门委员会在海平面变化的收支因素中没有考虑这一潜在的大型储水项。它对 barystatic 海平面上升的收支影响需要在十年尺度范围内才能达到平衡。

我们评估了在 12 年间(从 2002 年至 2014 年)陆地水储存量在海平面上升中的作用。我们利用卫星测量了时变重力数据,检验了由时变重力数据得到的地表质量的全球变化值。这些变化值可以很好地反映全球储水量的变化。从这些数据中,我们提取了陆地部分的储水量对海平面上升的净贡献的基于观测的预估数据。结合最近调整后的对冰川损失的估计值和最近对全球地下水消耗的估计值,我们能够将这种净质量的变化分解为冰川的贡献,直接受人类活动影响的陆地水储存的贡献,以及受气候影响的陆地水储存的贡献。

时变重力的测量结果由 NASA 的重力恢复和气候试验(GRACE)卫星观测得到。GRACE 提供了对地球重力场变化的月度观测值。这些变化是在去除了地球固体和大气变化产生的信号之后,在特定的时间和空间尺度范围内,由水和冰在地球系统中的运动造成的。自 2002 年 4 月以来,GRACE 提供了测量月度重力场的解决方案,并证明其是观测冰盖、冰川、雪块、区域地下水和地表水储存量变化的有效工具。先前的研究结果表明,GRACE 的测量精度通常随着观测域区的增大而增加,因此,GRACE 的观测结果可能有助于说明水文因素对海平面变化的影响,尽管只有 Rietbroek 等人曾经试图按照过程来分解各种因素的贡献。GRACE 提供的观测记录在不断地增长。结合最近的一些技术改进,包括在卫星间距离变化率的测量以及在模拟地球固体位移(补充材料、材料及方法)的改变导致的重力变化方面的改进,GRACE 的观测记录对于研究陆地水储存量的变化对海平面上升的影响将更加重要。

Text 12　Ancient Geodynamics and Global-Scale Hydrology on Mars

The western hemisphere of Mars is dominated by the Tharsis rise, a broad elevated (~10 km) region extending over 30 million square kilometers. Tharsis is the locus of large-scale volcanism and pervasive fracturing that resulted from the loading of the lithosphere, or outer elastic shell, by voluminous extrusive and intrusive magmatic deposits. Here we use recently acquired gravity and topography data from the Mars Global Surveyor (MGS) spacecraft to determine the effect of the mass load of Tharsis on the shape and gravity field of the rest of the planet. We test the hypothesis that the deformational response to the Tharsis load is responsible for the topographic trough and the heretofore unexplained ring of negative gravity anomalies (Figs. 12-1 and 12-2A) that surround the Tharsis rise, as well as for the major gravity and topographic highs that are antipodal to Tharsis. We examine the influence that Tharsis may have had on the timing, orientation, and location of fluvial features on the planet. Because of the enormous mass of the Tharsis load, understanding the global history of Mars requires understanding the role that Tharsis played in that history; the formation of Tharsis may have been an exceptional phenomenon in the evolution of the terrestrial planets.

To test the hypothesis that Mars displays a global deformational response to Tharsis loading, we use a spherical harmonic model of the loading of a spherical elastic shell and isolate the spatially variable Tharsis topography (Fig. 12-3A) as the only load on the planet. The degree of compensation of the load is about 95%, an assumption that produces consistency between predicted and observed topography but overpredicts the magnitudes of the associated gravity anomalies. Because we are interested primarily in the spatial correlation of models to observed fields, rather than model amplitudes, the mismatch of predicted and observed gravity is secondary to this discussion.

The locations of the observed and modeled ring of negative gravity anomalies (Fig. 12-2, A and B) around Tharsis, expanded to spherical harmonic order and degree 10 ($l = 10$), are consistent and include several of the intermediate-wavelength features within the ring (such as relative lows to the northwest, northeast, and east, and the relative high to the south). The model also predicts a topographic trough (here termed the Tharsis trough) around Tharsis (Fig. 12-3C). Such a topographic depression does surround Tharsis over at least 270° of azimuth (Fig. 12-3A). To the east of Tharsis, this trough extends northward from the Argyre impact basin, through Chryse and Acidalia Planitiae, to the North Polar basin. To the northwest of Tharsis, Arcadia and Amazonis Planitiae comprise the depression, whereas southwest of Tharsis, the depression becomes less evident. The fact that the ring of negative gravity anomalies persists to the southwest of Tharsis indicates that more than the surface topography is required to explain the gravity data here; one possibility is that the trough has been filled with sediments that are lower in density than average crustal material.

Antipodal to Tharsis, the model predicts a topographic high (here termed the Arabia bulge) over the elevated Arabia Terra (Fig. 12-3D); the predicted high also extends over the Utopia basin to the north. The planetary topographic dichotomy of a northern

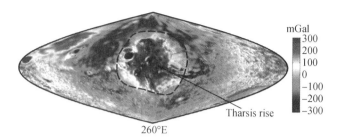

260°E

Fig. 12-1 Gravity anomaly image draped over a three-dimensional (3D) view of topography centered on Tharsis at 260°E longitude. This and all subsequent figures are in sinusoidal projection. The gravity anomaly image has been saturated at 6,300 milligals (mGal) and is expanded to $l = 60$. The prominent topographic feature in the center is the Tharsis rise and its volcanic constructs; the approximate boundary to Tharsis is shown as a dashed line. Valles Marineris extend eastward from Tharsis. Negative portions of the gravity anomaly field form a ring around Tharsis. A region antipodal to Tharsis (seen at the right and left sides of the map near 80°E longitude) is centered on Arabia Terra, which is both a topographic and a gravitational high.

lowland and a southern upland is not an element of a Tharsis loading model; however, the anomalously broad western rim of the Utopia basin (Fig. 12 - 3B) may be an expression of uplift resulting from Tharsis loading. The Hellas impact basin is also outside the realm of the model, yet both the modeled and observed topography show a downward slope in Hellas rim topography toward the South Pole. This slope, in the direction of Tharsis, suggests that a portion of the Hellas rim underwent vertical motion during the formation of the trough created by the Tharsis load. In both the observed and modeled gravity anomaly fields ($l = 10$), there is a gradient across Hellas and a high over Arabia Terra that extends northward to the Utopia basin, mirroring features in the long-wavelength topography (Fig. 12 - 2, C and D). The modeled gravity clearly does not account for the smaller scale, positive gravity anomalies at Elysium and the central Utopia basin, which reflect subsurface structures characterized, respectively, by volcanically thickened crust and a combination of crustal thinning during basin formation and infill of the basin depression. However, we conclude that the long-wavelength, nonhydrostatic gravity field of Mars is explained simply by the Tharsis load and the resulting global deformation of the lithosphere. Further, the shape of Mars is determined by these two quantities plus the northward pole-to-pole slope that formed in earliest martian history.

Extensional structures radial to the Tharsis rise and compressional structures generally concentric to the rise constitute the majority of the tectonic features in the Tharsis region. About half of these features are Noachian in age, suggesting that tectonic activity peaked early and decreased with time. The positions and orientations of both types of structures are matched by elastic shell loading models constrained by current gravity and topography fields. Successful models are able to predict strain levels

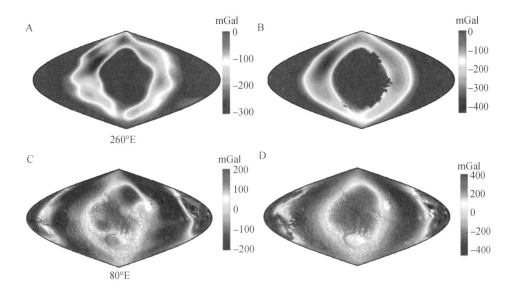

Fig. 12-2 (A) Observed and (B) modeled negative gravity anomaly ring around Tharsis, centered at 260°E longitude. The irregular boundary around Tharsis in the model results from defining the load boundary with a quarter-degree topographic grid (10). (C) Observed and (D) modeled gravity anomalies draped over a 3D view of observed topography centered on Arabia Terra at 80°E longitude. The observed gravity anomaly over Tharsis is shown in the model image. Different scales for observed and modeled fields result from an overprediction of model gravity. For (A) through (D), gravity fields are expanded to l=10 independently inside and outside of the load boundary.

comparable to those observed in Noachian structures and require that the extent of the load in the Noachian be comparable to that at present. Thus, the overall Tharsis load must have been largely in place by the Late Noachian. Because they are the direct response to the Tharsis load, the Tharsis trough and Arabia bulge must also have existed since Noachian time.

The development of the Tharsis trough and Arabia bulge thus should have influenced the location and orientation of martian valley networks and outflow channels. Valley networks are the most common drainage systems on Mars. Their similarity to terrestrial river systems suggests that the genesis of valley networks involved fluvial erosion, although the style of this erosion (such as surface runoff, groundwater discharge, or sapping) remains controversial. Valley network systems are confined mainly to the southern highlands on Noachian terrain and display variability in the number of tributaries, stream order, and planimetric form. Additionally, nearly all martian outflow channels originate in or flow into the Tharsis through (Fig. 12-3A).

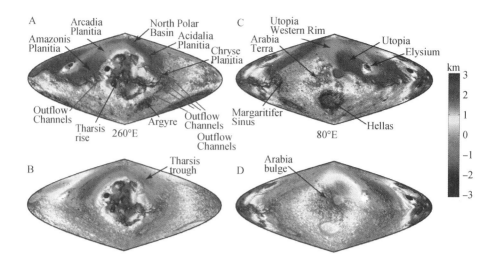

Fig. 12 - 3 Observed martian topography displayed for（A）the Tharsis and（B）the anti-Tharsis hemispheres compared with modeled topography（to $l=120$）for（C）the Tharsis and（D）the anti-Tharsis hemispheres. For the model, actual topography is shown in the Tharsis region. All figures are draped over a 3D view of shaded relief.

（Cited from Phillips R J，Zuber M T，Solomon S C，et al. Ancient geodynamics and global-scale hydrology on Mars［J］. Science，2001，291(5513)：2587-2591.）

New Words and Expressions

lithosphere	*n*.	岩石圈,陆界
anomalies	*n*.	异常现象
Tharsis rise		塔尔西斯隆起
topography	*n*.	地貌,地形学
fluvial	*adj*.	河的,河流的
terrestrial	*adj*.	陆地的,地球的
crustal	*adj*.	地壳的
Utopia	*n*.	乌托邦
Hellas	*n*.	希腊文(Greece)名称

译文:火星上古老的地质动力学与全球尺度的水文学

火星西半球的主体部分是塔尔西斯隆起,这是一个宽阔的隆起(～10公里)区域,延伸超过3 000万平方千米。塔尔西斯存在大规模的火山活动和广泛分布的断裂构造。这些断裂构造是由于大量喷出和侵入性的岩浆矿床作用于岩石圈或外弹性壳而形成的。我们利用最近从火星全球探测器(MGS)获得的重力和地形数据,确定了塔尔西斯的质量荷载对火星其余部分的形状和重力场的影响。我们验证了这样一个假设,即对塔尔西斯荷载的变形响应是产生围绕塔尔西斯隆起的沟槽和至今无法解释的负重力异常环带的原因(图12-1、12-2A);此外,主要的重力效应和与塔尔西斯相对的高地也是由其造成的。我们还研究了塔尔西斯隆起对火星上河流特征的时间、走向及位置可能造成的影响。由于塔尔西斯产生的荷载非常巨大,因此,要掌握火星的全球历史就需要了解塔尔西斯隆起在那段历史中发挥的作用;塔尔西斯隆起的形成可能是陆地行星进化过程中的一个特殊现象。

为了验证火星对塔尔西斯荷载表现出全球性的变形响应的假设,我们使用了对球形弹性壳体进行加载的球谐模型,并将空间可变的塔尔西斯地形(图12-3A)隔离为星球上的唯一载荷。荷载的补偿程度假设为大约95%。这一假设使得对地形的预测与观测结果保持一致,但过高预测了与之相关的重力异常的大小。由于我们感兴趣的主要是模型与观测场的空间相关性,而不是模型的振幅,因此在本次讨论中,重力预测值与观测值之间的不一致性是次要的。

当球谐阶次扩展到10时($l = 10$),通过观测和模拟得到的负重力异常环带(图12-2,A和B)在塔尔西斯隆起周边的分布位置是一致的。此外,环带内的若干中波的波长特性也能保持一致(例如西北部、东北部及东部的相对低点,以及南部的相对高点)。该模型还预测了塔尔西斯隆起周围的一条沟槽(这里称为塔尔西斯沟槽)(图12-3C),这些地形凹陷在至少270°的方位角内环绕着塔尔西斯(图12-3A)。在塔尔西斯隆起以东,这些沟槽从阿盖尔撞击盆地向北延伸,穿过克里斯和普兰尼提亚平原,到达北极盆地。在塔尔西斯隆起的西北部,阿卡迪亚平原和亚马孙平原构成了凹陷,而位于塔尔西斯隆起的西南部的凹陷则很不明显。负重力异常环带在塔尔西斯隆起的西南部持续存在的事实,表明除了地表地形以外还需要更多的资料才能解释这里的重力数据;一种可能性是沟槽中充满了密度比地壳物质的平均值更低的沉积物。

在与塔尔西斯隆起相反的一侧,该模型预测了位于阿拉伯台地上的一处高地(这里称为阿拉伯隆起)(图12-3D);预测到的高地也延伸到乌托邦盆地的北部。将行星地形一分为二的北部低地和南部高地并不是构成塔尔西斯荷载模型的一个要素,但乌托邦盆地异常宽广的西缘(图12-3B)可能是塔尔西斯荷载引起地表抬升的一种表现。海拉斯撞击盆地也在模型的预测范围之外,然而由模拟和观测得到的地形都显示海拉斯盆地边缘的地形朝着南极向下倾斜。这个朝着塔尔西斯方向倾斜的坡地,表明在塔尔西斯荷载引起的沟槽的形成期间,一部分海拉斯盆地的边缘区域经历了垂直运动。由观测和模拟得到的异常重力场($l = 10$)在海拉斯盆地和向北延伸至乌托邦盆地的阿拉伯台地上方都存在着梯度变化,这反映出长波地形的镜像特征(图12-2,C和D)。模拟得到的重力显然不能解释在伊利希尔和中央乌托邦盆地出现的小尺度、正重力异常现象。这些现象反映出的地下结构特征分别是火山作用导致的地壳增厚现象,以及在盆地形成期间出现的地壳减薄和盆地凹陷填充的组合现象。然而,我们得到的结论表明,火星的长波长、非流体静重力场可以简单地通过塔尔西斯荷载及由此产生的岩石圈的整体变形来解释。此外,火星的形状是由这两个量以及在火星早期历史中形成的向北倾斜且跨越两极的斜地所决定的。

塔尔西斯地区的主要构造特征是向着塔尔西斯隆起的径向伸展构造,以及与隆起大致同心的压缩构造。这些构造特征大约有一半起源于诺亚纪年代,这表明构造活动在早期就达到了高峰,然后随着时间的推移而减少。这两种结构的位置和方向均和受重力场和地形场约束的弹性壳体的加载模型相吻合。成功的模型能够预测与观测到的诺亚纪结构相当的应变量级,并且要求诺亚纪时期的荷载范围应与现在的荷载范围相一致。因此,塔尔西斯的总体荷载必须在晚诺亚纪之前就大致形成了。由于塔尔西斯沟槽和阿拉伯隆起是对塔尔西斯荷载的直接响应,所以它们也一定是自诺亚纪以来就存在的。

塔尔西斯沟槽和阿拉伯隆起的发展也应当会影响火星河谷网络及外流通道的位置与走向。河谷网络是火星上最常见的排水系统。它们与地球上的河流系统的相似性表明,火星河谷网络的形成也涉及河流的侵蚀现象,然而关于这种侵蚀(如地表径流、地下水排出或逐渐削弱)的形式仍然存在着争议。河谷网络系统主要局限于诺亚纪地形上的南部高地范围内,并在支流数量、流序和平面形式上存在着可变性。此外,几乎所有的火星河谷的外流通道都起源于塔尔西斯隆起,或是流入塔尔西斯隆起(图12-3A)。

Text 13　Human-Induced Changes in the Hydrology of the Western United States

Water is perhaps the most precious natural commodity in the western United States. Numerous studies indicate the hydrology of this region is changing in ways that will have a negative impact on the region. Between 1950 and 1999 there was a shift in the character of mountain precipitation, with more winter precipitation falling as rain instead of snow, earlier snow melt, and associated changes in river flow. In the latter case, the river flow experiences relative increases in the spring and relative decreases in the summer months. These effects go along with a warming over most of the region that has exacerbated these drier summer conditions.

The west naturally undergoes multidecadal fluctuations between wet and dry periods. If drying from natural climate variability is the cause of the current changes, a subsequent wet period will likely restore the hydrological cycle to its former state. But global and regional climate models forced by anthropogenic pollutants suggest that human influences could have caused the shifts in hydrology. If so, these changes are highly likely to accelerate, making modifications to the water infrastructure of the western United States a virtual necessity.

Here, we demonstrate statistically that the majority of the observed low-frequency changes in the hydrological cycle (river flow, temperature, and snow pack) over the western United States from 1950 to 1999 are due to human caused climate changes from greenhouse gases and aerosols. This result is obtained by evaluating a combination of global climate and regional hydrologic models, together with sophisticated data analysis. We use a multivariable detection and attribution (D & A) methodology to show that the simultaneous hydroclimatic changes observed already differ significantly in length and strength from trends expected as a result of natural variability (detection) and differ in the specific ways expected of human-induced effects (attribution). Focusing on the hydrological cycle allows us to assess the origins of the most relevant climate change impacts in this water limited region.

We investigated simultaneous changes from 1950 to 1999 in snow pack (snow water equivalent or SWE), the timing of runoff of the major western rivers, and average January through March daily minimum temperature (JFM T_{min},) in the mountainous regions of the western United States. These three variates arguably are among the most important metrics of the western hydrological cycle. By using the multivariable approach, we obtain a greater signal-to-noise (S/N) ratio than from univariate D&A alone (see below).

The SWE data are normalized by October-to-March precipitation (P) to reduce variability from heavy- or light-precipitation years. Observed SWE/P and temperature were averaged over each of nine western mountainous regions (Fig. 13-1) to reduce small-spatial-scale weather noise. The river flow variate is the center of timing (CT), the day of the year on which one-half of the total water flow for the year has occurred, computed from naturalized flow in the Columbia, Colorado, and Sacramento/San Joaquin rivers. CT tends to decrease with warming because of earlier spring melting.

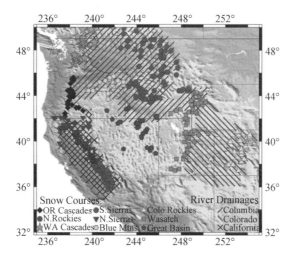

Fig. 13-1 Map showing averaging regions over which SWE/P and JFM T_{min} were determined. The hatching shows the approximate outline of the three main drainage basins used in this study.

Selected observations from these regions and variables are displayed in Fig. 13-2, showing the trends noted above, along with substantial regional differences and "weather noise." SWE/P trends in the nine regions vary from -2.4 to -7.9% per decade, except in the southern Sierra Nevada where the trend is slightly positive. The JFM T_{min}, trends are all positive and range from $0.28°$ to $0.43°C$ per decade, whereas the river CT arrives from 0.3 to 1.7 days per decade earlier. The challenge in D&A analysis is to determine whether a specific, predetermined signal representing the response to external forcing is present in these observations.

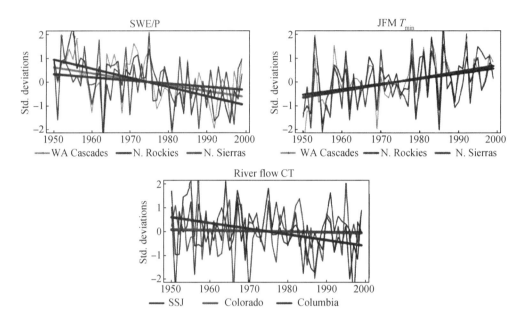

Fig. 13-2 Observed time series of selected variables (expressed as unit normal deviates) used in the multivariate detection and attribution analysis. Taken in isolation, seven of nine SWE/P, seven of nine JFM T_{min}, and one of the three river flow variables have statistically significant trends.

(Cited from Barnett T P, Pierce D W, Hidalgo H G, et al. Human-induced changes in the hydrology of the western United States [J]. Science, 2008, 319(5866): 1080-1083.)

New Words and Expressions

hydrological	*adj.*	水文学的
multivariable	*adj.*	多变量的
detection	*n.*	侦察,发掘
attribution	*n.*	归因,属性
perturbation	*n.*	扰动,摄动
exacerbate	*v.*	加重,使恶化
anthropogenic	*adj.*	人为的
arguably	*adv.*	可论证地,可争辩地

univariate	*adj*.	单变量的
precipitation	*n*.	降水
spatial	*adj*.	空间的
predetermined	*adj*.	预先决定的

译文:人为因素诱发的美国西部水文变化

水可能是美国西部最珍贵的天然物品。大量的研究表明,美国西部地区的水文正在以不利于该地区的方式发生变化。从1950年至1999年,山区的降水特征发生了变化,更多的冬季降水表现为雨水而不是雪水,而且雪的融化期也提前了,因此河流的径流也发生了相应的变化。在后一种情况下,河流的流量在春季相对增加,而在夏季月份则相对减少。这些影响还伴随着大部分地区的暖化效应,使夏季气候变得更加干燥。

西部一直存在着以数十年为周期的干旱期和湿润期交替的自然波动。如果当前的水文变化是由自然气候波动引起的干旱造成的,那么随后的湿润期将可能使水文循环恢复到原来的状态。但是根据全球和区域气候对人为污染物响应的模型,人类的影响可能会导致水文环境的变化。如果是这样,这些变化极有可能会加速发展,使得美国西部的水利基础设施的改造成为实际的需要。

在此,我们基于统计数据表明,从1950年至1999年,在美国西部观测到的水文循环(河流、温度和积雪)中的大多数低频变化是由人类释放的温室气体和悬浮微粒引起的气候变化造成的。这一结论是通过对全球气候和区域水文模型的组合进行评估,以及复杂的数据分析后获得的。我们使用多变量检测和归因(D&A)方法来展示,已经观测到的水文与气候的同步变化在长短和强度上显著不同于自然波动引起的预期变化趋势,而其与人为因素作用引起的预期变化趋势仅在某些具体方式上有所不同。关注水文循环使我们能够评估在这个水资源有限的地区气候发生变化的最重要的起因。

我们调查了1950年至1999年美国西部山区的积雪(雪水当量或称为SWE)、主要西部河流的径流时间和1月至3月间平均日最低气温(JFM T_{min})的变化。这三个变量可以说是西部水文循环的最重要的指标。通过多变量方法,我们获得了比单变量D&A方法更大的信噪比(见下图)。

我们利用10月至3月的降水(P)数据对SWE数据进行了归一化处理,以减少多雨年或少雨年导致的变异性。为减少小尺度空间的天气噪声,我们对西部9个山区的SWE/P和温度进行了平均(见图13-1)。我们采用的河流径流的变量是时间中心(CT)。它指的是一年中总流量的一半发生的日子,可以通过计算哥伦比亚、科罗拉多和萨克拉门托/圣华金河流的归化流量来确定。由于早春融雪,CT会随温度升高而降低。

从这些区域和变量中选择的部分观测结果显示在图13-2中。它们体现了上文中提到的变化趋势,以及显著的区域差异和"天气噪声"。9个地区的SWE/P趋势每十年在-2.4%到-7.9%之间变化,但内华达山脉南部的趋势略好。JFM T_{min}的变化趋势都是良好的,变化范围从每10年0.28°到0.43°C,而河流的时间中心CT每10年提早0.3天至1.7天到达。在D&A分析中面临的挑战是确定在这些观察中是否存在一个特定的预设信号来代表对外部作用的反应。

Text 14 Hydrological Cycle

The hydrological cycle is the most important carrier of water, energy, and matter (chemicals, biological material, sediments, etc), locally and globally (Fig. 14-1). The hydrological cycle acts like an enormous global pump that is driven mainly by two forces; solar energy and gravitation pull. Humans have ingeniously utilized this global and free pump to get irrigation water and to draw power from the enormous amount of energy that this cycle represents.

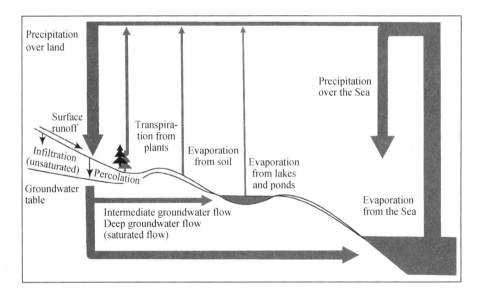

Fig. 14-1 The hydrological cycle (after Bonnier World map, 1975)

The incoming solar energy forces water to evaporate from both land and sea. Much of this vapour condensates and falls directly over the sea surface again (globally about 7/8 of the rainwater falls over the oceans). The remainder of the rainwater falls over land (globally about 1/8), and it falls as precipitation (rainfall, snow, and/or hail). This forms runoff as creeks, rivers, and lakes on the soil surface. A major part, however, infiltrates through the soil surface and forms soil water (water in the upper soil layers above the groundwater table, also called the unsaturated zone) that may later percolate (deeper infiltration) down to the groundwater (groundwater zone also called the saturated zone) level.

In the ground, water can also be taken up by plant roots and evaporate into the atmosphere through transpiration (evaporation through the plant leaves by pant respiration) or by direct evaporation from the soil. The total evaporation from both soil and plants is called evapotranspiration.

Carrier for Pullutants

The global cycle of water transports different types of chemical, biological, and sediment matter. Finally, these may be deposited in the sea because this is the lowest point in the system. If the release point for these constituents is known, it is often possible to predict the transport path by studying the local hydrology in the area. This is due to the fact that the pollutant often follows the same path as the water. However, chemical and/or biological transformation may also affect the pollutant.

Humans influence and change the general hydrological cycle to a great extent. Activities in the landscape directly affect the different components of the hydrological cycle. The chemical content of different hydrological parts is also increasingly affected by various activities such as industry, agriculture, and city life. Yet, the total amount of water on earth is constant. Water is neither created nor is disappearing from earth. However, the content of various biological and chemical elements can fluctuate, depending on the location of the hydrological cycle (Fig. 14-2).

Turnover time

Only a fraction of the total water volume is fresh water (about 2.7%). And, a major part (2/3) of this fresh water is located around the poles as ice and glaciers. The total amount of fresh water resources is consequently limited, and desalination of seawater is still an expensive process. The water contained in different components (shown in Fig.14-2) is continuously exchanged due to the constant movement of water.

The theoretical turnover time indicates the average time that it takes for the water volume to be exchanged once. For some components, e. g., water in rivers and atmosphere, the turnover time is very short, about one week (Table 14-1).This also indicates the theoretical transport time for pollutants released in various parts of the water cycle. A pollutant that accumulates on the glacial ice would theoretically surface again after about 8,000 years (Table 14-1). A pollutant released in the atmosphere or a river would be flushed out after about a week. However, the average turnover is theoretical, and assumes that the pollutants are not adsorbed or transformed by biological and geological media through which they are transported. In any case, the turnover time can give a rough estimation in order to understand transport velocities in different parts of the hydrological cycle.

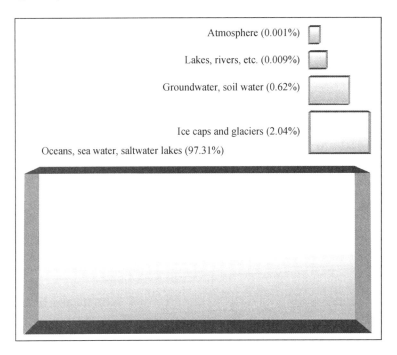

Fig. 14-2 The global distribution of water in different hydrological parts (after Bonnier World map, 1975)

Table 14-1 Average turnover time for water in different hydrological parts

Hydrological part	Volume (10⁶ km³)	%	Turnover time (year)
Oceans	1 370	94.2	3,000
Groundwater	60	4.1	5,000
Ice caps and glaciers	24	1.7	8,000
Lakes	0.3	0.02	10
Soil water	0.1	0.006	1
Atmosphere	0.01	0.001	1 week
Surface water	0.001	0.0001	1 week

Consequently，the turnover time may give a rough but general idea of how quickly a water particle may travel through a water body. This can be compared to a more detailed picture as seen in Table 14 - 2. The table shows typical velocities by which a water molecule or a pollutant travelling with the same speed as water may travel. Note that these values are based on approximation. Large variations may be expected，depending on the hydrological situation，the hydraulic conductivity of the geologic medium，etc.

Table 14-2 Typical average velocity for a water molecule

Water body	Soil type	Water molecule velocity
Soil water (vertical)		1-3 m/year
Groundwater	moraine close to soil surface	1-10 m/day
	1 m depth	0.1 m/day
	deep	0.01 m/day
	cracked bed rock	0-10 m/day
Creek		0.1 m/s≈10 km/day
River		1 m/s≈100 km/day

Humans interact with the natural hydrological system by diverting water for different activities. A major part of this water, about 80% on a global scale, is used for irrigation and production of agricultural products. The remaining 20% is used for industrial needs and domestic water supply. Figure 14-3 shows, in a schematic way, how water is used in households. It is seen that for a rich and developed country, a very small part (about 3%) is used for direct consumption (drinking water and cooking). The remaining part is used for sanitation purpose. For a poor country in the developing world, average per capita water consumption may be only 20 liters a day. The UN recommends that people need a minimum of 50 liters of water a day for drinking, washing, cooking, and sanitation.

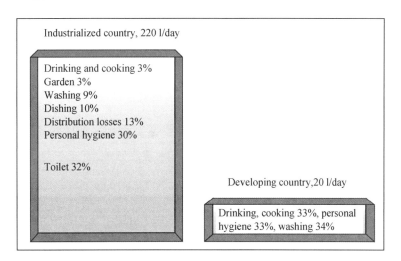

Fig. 14-3 Domestic water consumption (after Bonnier World map, 1975)

(Cited from Ojha C, Berndtsson R, Bhunya P. Engineering hydrology [M]. 1st ed. Oxford: Oxford University Press, 2008.)

New Words and Expressions

hydrological cycle		水文循环
ingeniously	adv.	贤明地,有才能地
evaporate	vt.	使蒸发,使脱水
condensate	n.	冷凝物,固化物
infiltrate	v.	浸润
evapotranspiration	n.	(气象)蒸散;土质水分蒸发蒸腾损失总量
constituent	n.	成分
component	n.	部件,组件,成份
divert	v.	使转移
schematic	adj.	图解的;概要的
irrigation water		灌溉用水
utilize	v.	利用
vapour	n.	蒸气(同 vapor),水蒸气
runoff	n.	(水文)流走的东西
percolate	vi.	过滤,渗出,浸透
landscape	n.	风景,风景画,景色
turnover time		周转时间

译文:水文循环

水文循环是水、能量和物质(化学物质、生物物质、沉积物等)在区域和全球范围内最重要的载体(图 14-1)。水文循环就像一个巨大的全球泵,主要由两种力量驱动:太阳能和重力牵引。人类巧妙地利用了这个全球性的免费水泵来获取灌溉用水,并从这个循环所蕴藏的巨大能源中获取能量。

从太阳获得的能量使水从陆地和海洋蒸发。大量水蒸气再次冷凝并直接落在海面上(全球约有 7/8 的雨水落在海洋上)。其余部分的雨水落在陆地上(全球约占 1/8),形成了降水(降雨、降雪或冰雹),由此产生了地表径流,如小溪、河流和湖泊。然而,大部分的降水会从地表渗入地下并形成土壤水(位于地下水位以上的上部土层中的水,也称为非饱和带),随后可能继续下渗(深层渗透)到达地下水位(地下水区域,也称为饱和带)。

在地下,水也可以被植物根系吸收,然后通过蒸腾作用(通过植物叶片的呼吸蒸发)蒸发到大气中,或直接从土中蒸发到大气中。土和植物的总蒸发量被称为蒸散量。

污染物的搬运者

水的全球循环可以传输不同类型的化学、生物和沉积物质。最后,这些物质可能沉积在海洋中,因为这是水循环系统中的最低点。如果已知这些成分的排放点,一般就可以通过研究该地区的水文来预测运输路径。这是因为污染物通常沿着与水相同的路径进行传输。然而,化学和/或生物转化也可能对污染物的传输造成影响。

人类活动在很大程度上影响和改变了整个水文循环。人在自然环境中的活动直接影响了水文循环的各个组成部分。不同水文部分的化学含量也日益受到各种人类活动的影响,如工业、农业和城市生活。不过,地球上的总水量是恒定的,因为地球上既没有水产生,也没有水消失。然而,根据水文循环地理位置,各种生物和化学元素的含量会有一定的波动(图14-2)。

周转时间

淡水仅占总水量的一小部分(约2.7%)。而且,这些淡水的大部分(2/3)以冰和冰川的形式位于两极附近。因此,淡水资源的总量是有限的,而且海水淡化依然成本高昂。由于水的持续运动,不同组分的水(图14-2)一直在不断地发生交换。

周转时间在理论上表示水量交换一次需要的平均时间。某些组分中的水,例如在河流和大气中的水,周转时间很短,约为一周(表14-1)。这个时间也可以作为在水循环各个环节中释放出的污染物的理论传输时间。从理论上讲,积聚在冰川中的污染物要经过大约8 000年才能再次露出水面(表14-1)。在大气或河流中释放出的污染物在一个星期后就会被冲走。然而,平均周转时间仅是理论上的数值,并假设污染物不会被生物和地质介质吸附或转化。总之,周转时间可以给出一个粗略的估计,以便了解污染物在水文循环的不同环节的传输速度。

因此,周转时间可以提供一个粗略但普遍的概念,即水粒子在水体中可以流动得多快。这可以与表14-2所示的一组更详细的数据进行比较。该表显示了水分子或污染物的典型运动速度与水的运动速度相同。但需要注意的是,这些数值是基于近似估算得到的。如果水文环境、地质介质的导水率等情况发生了变化,结果可能会出现较大的差别。

人类需要利用水进行不同的生产活动,从而与自然水文系统发生了相互作用。这些水的很大一部分——在全球范围内约占80%——被用于灌溉和农业生产。剩下的20%则是工业用水和生活用水。图14-3以示意图的方式展示了家庭用水情况。可见,对于富裕的发达国家来说,只有很小一部分(约3%)用于直接消费(饮用和烹饪),其余部分用于卫生目的;而对于贫穷的发展中国家来说,人均用水量可能只有每天20升。联合国建议每人每天至少需要50升水用于饮用、洗涤、烹饪和卫生。

Text 15　Land Subsidence Due to Groundwater Drawdown in Shanghai

Shanghai is located on the deltaic deposit of the Yangtze River, as illustrated in Fig. 15-1. Fig. 15 - 2 shows the main geological strata. The total thickness of the

Quaternary deposit is about 300 m (Shanghai Geology Office, 1976, 1979). Excessive pumping of the groundwater caused compression of the Quaternary deposit, and subsidence in Shanghai. Although several countermeasures, including recharge of the groundwater, have been adopted to mitigate the rate of subsidence, up to the present the recorded cumulative subsidence has been 2–3 m in the central area of Shanghai. This subsidence has caused a lot of social problems. The immediate problem is the increase in the possibility of flooding. From 1981 to 1994, rainfall flooding occurred 22 times, at a rate of almost twice per year (Liu, 2001). The possibility of tidal flooding also increased.

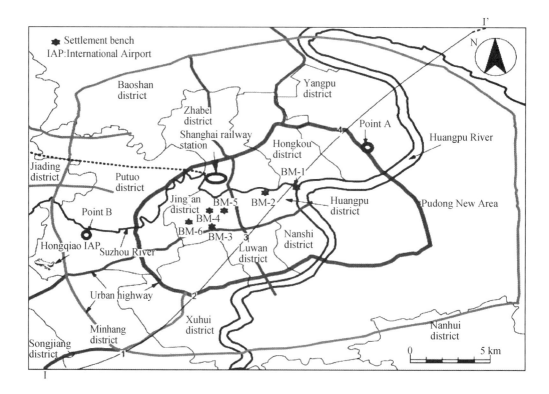

Fig. 15-1 Location of monitoring points and cross-section I—I'

CHAI, SHEN, ZHU AND ZHANG

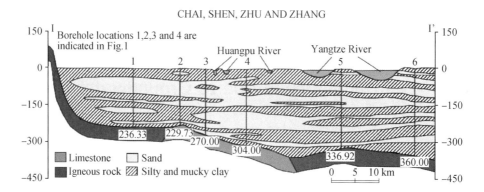

Fig. 15-2 Shanghai geological strata (cross-section I—I' in Fig. 15-1) (after Shanghai Geology Office, 1979)

From 1956, the height of the dike along the coast line was increased four times, with the crest elevation rising from 5 m to 6.8 m. Other problems caused by the land subsidence are damage to the sewerage system, road, buildings, and subway tunnels etc.

In Shanghai the monitoring of land subsidence started in 1921. The history of land subsidence in the urban area of Shanghai can be divided into two periods: namely a rapid subsidence period (1921—1965) and a controlled period (1965—present) (Zhang & Han, 2002). However, there has been a tendency for the subsidence rate to increase after 1990. In this paper, the mechanism of land subsidence due to excessive pumping of groundwater is discussed first. Then the measured field data are presented, and the relationship between groundwater pumping and land subsidence is discussed.

The most important mechanism causing land subsidence is excessive pumping of groundwater, which causes drawdown of the water table in the aquifer and a decrease in pore water pressure and an increase of effective stress in the ground. However, owing to the supply from rainfall and other surface water, the phreatic level in the surface soil layer does not change much. As a result, there will be a total head difference within the soil layers above the aquifer with pore water pressure drawdown, and there will be a water flow from the ground surface towards the aquifer. Therefore the consolidation caused by pore water pressure drawdown in an aquifer is different from that under a surcharge loading. The final state of the latter case is complete dissipation of excess pore pressure and a zero water flow rate. However, for the former case, the final state is a steady water flow. For multi-layer ground, after reaching a steady state, to satisfy the continuity condition of the flow, the following equation holds:

$$q = k_{v1} i_1 = k_{v2} i_2 = k_{vi} i_i = \cdots \tag{15-1}$$

where q is flow rate, k_{vi} is the vertical hydraulic conductivity of the ith layer, and i_i is the hydraulic gradient within the ith layer. It is obvious that, to satisfy equation (1), the layer with a lower k_v value should have a higher i value, and therefore an increased pore water pressure drop across the layer. For the two-layer case, there will be three possible patterns of pore water pressure distribution within the clay layers (see Fig. 15-3):

(a) pattern A: a triangular distribution that corresponds to a condition where the hydraulic conductivity (k_v) of the aquitard is uniform [see Fig. 15-3(a)]

(b) pattern B: a rightward concave distribution with a condition that the aquitard is layered and the k_v value of the upper layer is smaller than that of the lower layer ($k_{vu} < k_{vl}$) [Fig. 15-3(b)]

(c) pattern C: a leftward concave distribution with the condition ($k_{vu} > k_{vl}$) [Fig. 15-3(c)].

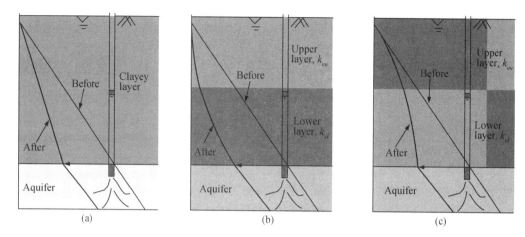

Fig. 15-3 Some possible pore water pressure drawdown patterns: (a) pattern A (uniform k_v);
(b) pattern B $k_{vu} < k_{vl}$); (c) pattern C ($k_{vu} > k_{vl}$)

As indicated in the figures, with the same pore water pressure drawdown in the aquifer, the effective stress increase in the aquitard is different: that is, pattern B [Fig. 15-3(b)] has the largest increase and pattern C [Fig. 15-3(c)] has the smallest increase, and so the amount of compression will be different. There are several field measurements that show a pore water pressure distribution like pattern C (e.g. Bergado et al., 1996). This is because, if the other properties of soil are the same, owing to the consolidation caused by the self-weight, the void ratio reduces with the increase of depth. Normally the k_v value is related to void ratio, e, and the smaller is e, the lower is the k_v value (e.g. Taylor, 1948). Vaughan (1994) discussed the reduction of hydraulic conductivity with increase of effective stress (reduction of void ratio) for the same soil and presented a few field-measured examples of pattern C [Fig. 15-3(c)], such as the pore water pressure distributions in the core of dams, which involved a downward steady water flow.

There are two possible 'mistakes' in evaluating the effective stress increase in the ground caused by pore water pressure drawdown in an aquifer. One is to consider the same amount of pore water pressure drawdown in the whole aquitard, and the other is to considering that Fig. 15-3(a) is universally applicable.

(Cited from Chai J C, Shen S L, Zhu H H, et al. Land subsidence due to groundwater drawdown in Shanghai [J]. Geotechnique, 2004, 54(2): 143-147.)

New Words and Expressions

Quaternary deposit		第四纪沉积层
countermeasure	*n*.	对策
cumulative	*adj*.	累积的

subsidence	*n.*	下沉,沉降
aquifer	*n.*	含水层
hydraulic conductivity		渗透系数
triangular	*adj.*	三角的
concave	*n.*	凹面
monitoring	*n.*	监视,监测
drawdown	*n.*	减少,水位降低
surcharge	*n.*	超载,额外费
hydraulic gradient		(流)水力梯度
rightward	*adj.*	向右的

译文:上海市地下水位下降引起的地面沉降

如图 15-1 所示,上海位于长江三角洲沉积区。图 15-2 则显示了上海的主要地层,其中第四纪沉积层的总厚度约 300 米(Shanghai Geology Office,1976,1979)。地下水的过量开采造成了上海第四纪地层的压缩和地面沉降。虽然上海市采取了包括地下水回灌在内的一些措施来减小沉降速率,但到目前为止,上海市中心城区记录的累计沉降量已达 2~3 米。地面沉降造成了许多社会问题,其中最紧迫的问题就是发生洪水灾害的可能性增加了。从1981 年到 1994 年,降雨引发的洪水灾害发生了 22 次,几乎每年两次(Liu,2001)。潮水引发洪水灾害的可能性也有所增加。从 1956 年起,沿海岸线的堤防高度增加了 4 倍,坝顶高程从 5 米上升到 6.8 米。地面沉降引起的其他问题还包括对排污系统、道路、建筑物、地铁隧道等的破坏。

上海地面沉降监测始于 1921 年。上海市区地面沉降的历史可以分为两个时期:一个是快速沉降时期(1921—1965 年),一个是沉降控制时期(1965 年至今)。然而,1990 年以后,地面沉降速率有增加的趋势。本文首先讨论了地下水超采引起地面沉降的机理,接着在此基础上,给出了现场实测数据,讨论了地下水开采与地面沉降的关系。

引起地面沉降的最重要的原因是地下水的过度开采,它导致含水层地下水位下降,孔隙水压力降低,土中有效应力增加。然而,由于降雨和其他地表水的补给,表层土中的潜水水位变化不大。因此,随着孔隙水压的下降,含水层上方的土层中存在着的总水头差,地下水会从地表流向含水层。然而,由含水层中孔隙水压下降引起的固结与超载作用下的固结是不同的。后一种情况的最终状态是超孔隙水压完全消散,且水流量变为零。然而,前一种情况的最终状态则是水流量达到稳定。对于多层土体来说,在达到稳定状态后,为了满足流动的连续性条件,下式必须成立:

$$q = k_{v1}i_1 = k_{v2}i_2 = k_{vi}i_i = \cdots \tag{15-1}$$

式中,q 为流量,k_{vi} 为第 i 层的垂直渗透系数,i_i 为第 i 层内的水力梯度。显然,为了满足式(15-1),k_v 值较低的地层应该具有较高的 i 值,从而增加了该地层的孔隙水压降幅。在双地

层的情况下,粘土层内部孔隙水压力的分布可能有三种模式(图 15-3):

a) 模式 A:如果含水层渗透系数不变,呈三角形分布[图 15-3(a)]。

b) 模式 B:如果存在弱透水层且上层的 k_v 值大于下层的 k_v 值($k_{vu} < k_{vl}$),呈右凹型分布[见图 15-3(b)]。

c) 模式 C:如果($k_{vu} > k_{vl}$),呈左凹型分布[图 15-3(c)]。

图中的结果表明,如果含水层中孔隙水压力下降相同的幅度,其有效应力增加的幅度就会不同。也就是说,模式 B 的有效应力增加的幅度最大,而模式 C 的有效应力增加的幅度最小,所以土层的压缩量将不同。有几种现场测量数据表明孔隙水压力的分布类似于模式 C(e.g. Bergado et al.,1996)。这是因为,如果土的其他性质相同,自重引起的固结会导致孔隙率随着深度的增加而减小。k_v 值通常与孔隙率 e 有关,e 越小,k_v 值就越低(e.g. Taylor,1948)。Vaughan(1994)讨论了对于同一种土,渗透系数随着有效应力的增加(孔隙率变小)而减小的情况,并展示了一些符合模式 C 的孔隙水压分布的现场实测例子,比如大坝心墙部位的孔隙水压分布,其中涉及下降的稳定水流。

在评价含水层中孔隙水压下降引起的地层有效地应力的增加时,存在两种可能的"错误":一种是认为整个含水层的孔隙水压的下降量相同,另一种则是认为图 15-3(a)的模式是普遍适用的。

Unit 4   Soil and Rock Mechanics

Text 16   Acute Sensitivity of Landslide Rates to Initial Soil Porosity

---

In popular metaphor, landslide processes begin spontaneously and gain momentum as they proceed, but what determines how real landslides move? Can small differences in initial conditions cause some landslides to accelerate catastrophically and others to creep intermittently downslope? The distinction is important because rapid landslides pose lethal threats, whereas slow landslides damage property but seldom cause fatalities.

A longstanding hypothesis holds that landslide behavior may depend on initial soil porosity, because soils approach specific critical-state porosities during shear deformation. Tests on small soil specimens indicate that dense soils (initially less porous than critical) dilate as they begin to shear, whereas loose soils (initially more porous than critical) contract. Dilation can reduce pore water pressures and thereby retard continued deformation by increasing normal stresses and frictional strength at grain contacts, whereas contraction can increase pore water pressures and thereby reduce frictional strength. Positive feedback between frictional strength reduction and soil contraction may cause some landslides to transform into liquefied high-speed flows.

To isolate the effect of initial soil porosity on landslide style and rate, we conducted large-scale experiments under closely controlled conditions. In each of nine landslide experiments, we placed a 65-cm-thick, 6-m^3 rectangular prism of loamy sand soil (Table 16-1) on a planar concrete bed inclined 31° from horizontal and bounded laterally by vertical concrete walls 2 m apart (Fig. 16-1). The downslope end of each soil prism was restrained by a rigid wall, which ensured that deformation occurred at least partly within the soil mass (rather than along the bed) and that land sliding included a rotational component.

Table 16-1. Mean physical properties of loamy sand used in landslide experiments. N denotes the number of samples on which measurements were made in each experiment or soil test, and 6 values indicate 1 SD from the mean

Property (method or definition)	Loose soil experiment	Dense soil experiment
Mean texture (weight %) (wet sieving and sedigraph)	89% sand, 6% silt, 5% clay ($N=8$)	89% sand, 6% silt, 5% clay ($N=8$)
Initial moist bulk density (g/cm³) [excavation method (15)]	1.44 ± 0.06 ($N=6$)	1.82 ± 0.03 ($N=6$)
Initial water content (water mass/solid mass)	0.12 ± 0.005 ($N=6$)	0.14 ± 0.007 ($N=6$)
Initial porosity (1−dry bulk density/2.7)	0.52 ± 0.02 ($N=6$)	0.41 ± 0.01 ($N=6$)
Hydraulic conductivity (cm/s) (permeameter tests*)	0.025 ± 0.007 ($N=3$)	0.0022 ± 0.00005 ($N=3$)
Hydraulic diffusivity (cm²/s) (drained compression tests*)	11 ± 4 ($N=2$)	28 ± 6 ($N=2$)
Friction angle at failure (degrees) (triaxial unloading tests*)	29 ± 2 ($N=2$)	41 ± 1 ($N=2$)

* These tests were conducted on reconstituted soil compacted to the desired porosity.

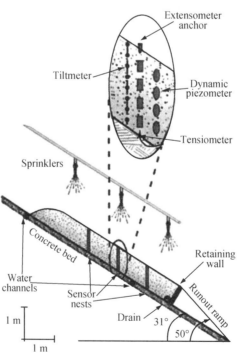

Fig. 16-1 Schematic longitudinal cross section of landslide experiments conducted at the U.S. Geological Survey debris flow flume, Oregon. The magnified ellipse depicts the positioning of sensors in vertical nests.

Different methods of soil placement yielded different initial porosities. The highest porosities (>0.5) were attained by dumping the soil in 0.5-m³ loads and raking it into position, without otherwise touching its surface. Lower porosities resulted from placing the soil in 10-cm layers parallel to the bed and compacting each layer with either foot traffic or 16-Hz mechanical vibrations that delivered impulsive loads of ~2 kPa at depths of 10 cm. After placement of each soil prism, we determined porosities by excavating four to nine~1 - kg samples at various depths and measuring their volumes, masses, and water contents. No systematic variations of porosity with depth were detected.

Our suite of landslide experiments included individual tests with initial porosities ranging from 0.39 ± 0.03 to 0.55 ± 0.01 (± 1 SD sampling error for an individual experiment). Ancillary tests of the same soil in a ring-shear device and triaxial cell produced dilative shear failure when initial porosity was $\leqslant 0.41$ and contractive shear failure when initial porosity was $\geqslant 0.46$ (Fig. 16-2 and Table 16-1). Landslides with initial porosities that bracketed the range from 0.41 to 0.46 were therefore of greatest interest.

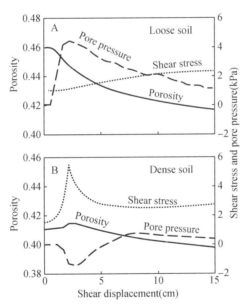

Fig. 16-2　Behavior of the loamy sand (Table 1) in a loose state (initialporosity 0.46) understate (initial porosity 0.41) when subjected to deformation in a ring-shear device. The device imposed shear displacements at 2 cm/s under constant normal loads of 10 kPa and permitted free drainage of water from the top and bottom of 7-cm-thick saturated soil specimens. Under these conditions, measured shear stress is a surrogate for effective soil strength. Loose soil (A) contracted monotonically during shearing, resulting in decreased porosity, transiently elevated pore pressure, and no peak in effective strength. Dense soil (B) initially dilated during shearing, resulting in increased porosity, transiently reduced pore pressure, and a prominent peak in effective strength. The porosity of the dense soil subsequently declined in response to breakage of soil aggregates. Triaxial unloading tests using a protocol described in (7) also produced contractive behavior for porosity 0.46 and dilative behavior for porosity 0.41.

Landslide motion was measured with two ground-surface extensometers and 17 or 18 subsurface tiltmeters arranged at depth increments of ～7 cm in two vertical nests. Pore water pressures were measured with 12 tensiometers and 12 dynamic piezometers arranged in three vertical nests at depth increments of ～20 cm (Fig. 16-1). Data from each sensor were logged digitally at 20 Hz for the duration of each experiment.

To induce landsliding, soil prisms were watered with surface sprinklers and through subsurface channels that introduced simulated groundwater (Fig. 16-1). Rising water tables were kept nearly parallel to the impermeable bed by adjusting discharge from a drain at the base of the retaining wall. Although preliminary experiments indicated that different styles and rates of water application influenced the onset of slope failure, this influence became negligible as failure occurred and instigated changes in soil porosity.

Landslides with differing porosities displayed sharply contrasting dynamics (compare Figs. 16-3 and 16-4). Each of four landslides with initial porosities>0.5 failed abruptly and accelerated within 1 s to speeds>1 m/s. The surfaces of these landslides appeared fluid and smooth, and data from dynamic piezometers confirmed that pore water

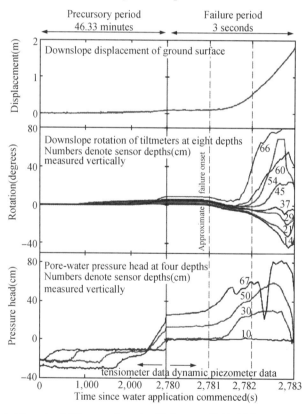

Fig. 16-3 Data recorded in a landslide experiment with loose soil (initial porosity 0.52±0.02). To reveal details of behavior during the 3-s failure period, the time axis is expanded 927 times. All sensors were initiallypositioned 2.3 m upslope from the retaining wall (ellipse in Fig. 16-1). Different sensors measured pore water pressure heads in the precursory and failure periods.

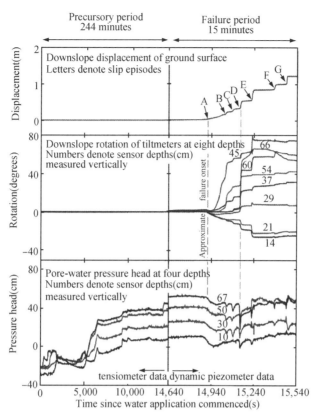

Fig. 16-4 **Data recorded in a landslide experiment with dense soil (initial porosity 0.41 ± 0.01). To reveal details of behavior during the 15 min failure period, the time axis is expanded 16 times. All sensors were initially positioned 2.3 m upslope from the retaining wall (ellipse in Fig. 16-1). Different sensors measured pore water pressure heads in the precursory and failure periods.**

pressures rose rapidly during failure and reached levels nearly sufficient to balance total normal stresses and liquefy the soil (Fig. 16-3). Three landslides with initial porosities indistinguishable from the critical porosity (0.44 ± 0.03, 0.44 ± 0.03, and 0.42 ± 0.03) displayed inconsistent behavior, including slow slumping of a single soil block, episodic slumping of multiple blocks, and moderately rapid (~0.1 m/s) slumping that ceased after <0.5 m displacement. Dynamic piezometer data from these experiments revealed a complex mix of dilative and contractive soil behavior during failure. The landslide with the lowest and least variable initial porosity (0.4 ± 0.01) displayed the clearest dilative soil behavior as it underwent slow episodic motion (Fig. 16-4). Our attempt to induce a landslide with still lower porosity (0.39 ± 0.03) ended uneventfully because we could not impart pore water pressures sufficient to trigger slope failure.

Figure 16-3 illustrates how land sliding of loose soil (initial porosity 0.52 ± 0.02) can lead to rapid acceleration in the course of only 1 s. After about 2,400 s of precursory sprinkling (with no groundwater inflow), positive pore water pressures developed first near the concrete bed and thereafter at shallower depths as a water table accreted vertically at rates ~ 0.05 cm/s. This wetting caused soil compaction，evidenced by a slight downslope rotation of tiltmeters at all depths，downslope surface displacement of nearly 10 cm，and vertical surface settlement of about 2 cm. As a consequence，average porosity declined to about 0.49，but the soil remained looser than critical. The soil developed no surface cracks or other visible signs of instability during this precursory period.

(Cited from Iverson R M，Reid M E，Iverson N R，et al. Acute sensitivity of landslide rates to initial soil porosity [J]. Science，2000，290(5491)：513-516.)

New Words and Expressions

acute	adj.	敏锐的
porosity	n.	孔隙率
episodes	n.	插曲,片断,一集
spontaneously	adv.	自然地,自发地
momentum	n.	势头;(物)动量,动力
retard	vt.	使减速,妨碍,阻止,推迟
concrete bed		混凝土床（底座）
rake	v.	耙,梳理,扫视,搜寻
extensometer	n.	伸缩计
instigate	v.	使(某事物)开始或发生,鼓动
oneset	n.	攻击,袭击,开始,动手
imperceptibly	adv.	极微地,微细地
dilated	adj.	膨胀的
metaphor	n.	象征,隐喻,暗喻
longstanding	adj.	长时间的,长期存在的
planar	adj.	平面的,平坦的
dump	v.	倾倒
ancillary	adj.	辅助的,附属的,补充的
tiltmeter	n.	倾斜仪,测斜仪

译文：滑坡速率对土的初始孔隙率的高度敏感性

当前的主流观点认为,滑坡过程是自发开始的,并且随着滑坡进程而得到动力,然而是什么因素决定了真实滑坡的运动方式呢？初始条件的微小差异是否会导致一些滑坡灾难性地加速下滑,而另一些滑坡则是间歇性地蠕动下滑? 这种区别很重要,因为快速滑坡会造成致命的威胁,而缓慢滑坡仅会造成财产损失而很少造成人员伤亡。

一种长期存在的假设认为,滑坡行为可能取决于土的初始孔隙率,因为土在剪切变形过程中会接近特定的临界状态孔隙率。对小土样的试验表明,致密土(初始孔隙率小于临界值)在开始剪切时膨胀,而松散土(初始孔隙率大于临界值)则会收缩。膨胀可以降低孔隙水压力,从而通过增加颗粒接触处的法向应力和摩擦强度来延缓持续的变形;而收缩可增加孔隙水压力,从而降低摩擦强度。摩擦强度降低与土体收缩之间的正反馈作用可能导致一些滑坡转化为液化高速流动。

为区分初始孔隙率对滑坡型式和滑坡速率的影响,通过精密控制变量,我们进行了大型模型试验。在九个滑坡试验中,我们将一个 65 厘米厚、6 立方米的壤质砂土矩形棱柱(表 16-1)置于平面混凝土床上。该平面混凝土床与水平方向呈 31 度角,横向由间隔两米的垂直混凝土墙约束(图 16-1)。每个土棱柱的下坡端都受到刚性墙的约束,这确保了变形至少部分发生在土体内部(而不是沿着混凝土床),并且滑动包括了旋转分量。

通过不同的制样方法可以获得不同的初始孔隙率。最大孔隙率(>0.5)是通过将土在0.5 立方米的载荷下倾倒并将其耙到位,此外没有再接触其表面。较低的孔隙率是将土平铺在混凝土床上,分层铺设,每层 10 厘米厚,并用脚踩或振动频率为 16 赫兹的机械振动压实每一层,从而在 10 厘米深度处产生约 2 千帕斯卡的冲击载荷。在放置每个土棱柱之后,我们在不同深度挖掘了 4~9 个 1 千克重的样品并测量它们的体积、质量和含水量,以此确定了样品的孔隙率。我们未发现孔隙率随深度呈现系统性的差异。

本组滑坡试验包括初始孔隙率为 0.39±0.03 至 0.55±0.01(±1SD 个体试验取样误差)的单体试验。辅助试验表明:当初始孔隙率小于等于 0.41 时,相同的土在环形剪切试验和三轴试验中会产生膨胀剪切破坏;当初始孔隙率大于等于 0.46 时产生收缩剪切破坏(图 16-2 和表 16-1)。因此,初始孔隙率介于 0.41 到 0.46 之间的滑坡最令人感兴趣。

滑坡运动由两个地表伸缩仪和 17 或 18 个安放在两个垂直孔穴中且间隔 7 厘米埋深的倾斜仪测量。孔隙水压力由 12 个张力计和 12 个动态压强计测量。这些传感器布置在 3 个垂直的孔穴中,相互之间埋深间隔 20 厘米(图 16-1)。在试验期间每个传感器的数据采集频率是 20 赫兹。

为了诱发滑坡,我们用表面喷洒器给土棱柱体洒水,并通过地下通道注水来模拟地下水的作用(图 16-1)。通过调整挡土墙底部排水沟的排水量,我们使上升的水位几乎与防渗的混凝土床平行。尽管初步试验表明,不同施水方式和施水速率对边坡失稳的发生有影响,但随着失稳的发生和土孔隙率的变化,这种影响变得微不足道。

不同孔隙率的滑坡表现出完全不同的动力特性(比较图 16-3 和图 16-4)。4 个初始孔隙率>0.5 的滑坡均突然发生破坏,并在 1 秒内加速至 1 米/秒以上,滑坡表面呈流体状、光滑状。动态水压力计的测量结果表明:孔隙水压力在破坏期间迅速上升,达到几乎足以平衡总法向应力和使土体液化的水平(图 16-3)。3 个初始孔隙率与临界孔隙率无明显差别的滑坡(0.44±0.03、0.44±0.03 和 0.42±0.03)表现出不一致性,包括单块土体缓慢崩塌,多块土体间歇崩塌以及经过小于 0.5 米位移后停止的中速崩塌(~0.1 米/秒)。这些试验中的动态孔压计数据揭示了土体破坏期间存在膨胀和收缩的复杂混合行为。具有最低和最小差异初始孔隙率(0.4 ± 0.01)的滑坡在经历缓慢的间歇性运动时表现出最明显的膨胀行为(图16-4)。我们试图诱发孔隙率较低(0.39 ± 0.03)的滑坡发生失稳,但失败了,因为我们不能提供足够的孔隙水压力来触发滑坡破坏。

图16-3说明了松散土体(初始孔隙率0.52 ± 0.02)的滑坡如何仅在1秒的时间内达到很高的加速度。在大约2 400秒的初期喷洒(没有地下水流入)之后,正孔隙水压力首先在混凝土床附近发展,然后在较浅的深度发展,地下水位在垂直方向以～0.05厘米/秒的速率增长。这种水分的增加导致土体压实,这可由倾斜仪在所有深度处轻微地朝下坡方向旋转,下坡表面位移接近10厘米和垂直表面沉降约2厘米来证明。这样一来,平均孔隙率下降到约0.49,但土体仍然比临界状态松散。在这一前兆期间,土体表面没有出现裂缝或其他明显的不稳定迹象。

Text 17　Simple Scaling of Catastrophic Landslide Dynamics

Seismic radiation from landslides is broad-band and complex. Short-period waves taking place within the granular mass and along its sliding boundary. They are distributed in time and low in amplitude compared with the impulsive radiation associated with the sudden stress drop in tectonic earthquakes. Long-period waves radiated by landslides are simpler: They are generated by the broad cycle of unloading and reloading of the solid Earth induced by the bulk acceleration and deceleration of the landslide mass. The corresponding momentum exchange is complicated by entrainment and deposition during motion and by topographic undulations along the slide path. Characteristic unloading reloading times in large landslides are several tens of seconds, making them efficient sources of seismic waves at periods of that order.

Traditional earthquake monitoring conducted by national and international agencies is designed for detection of impulsive short-period seismic waves and for location of associated tectonic earthquakes and explosions. Landslide detections are rare. A complementary method based on near real time data from the Global Seismographic Network (GSN) allows for the detection of seismic events through continuous back-projection of the long-period wavefield. This event detection algorithm detects>90% of magnitude $M \geqslant 5.0$ shallow earthquakes reported by other agencies and identifies about 10 events each month that are not in other seismicity catalogs. Some of these unassociated events have been correlated with large-scale glacier calving and volcanic unrest. Here, we identify and investigate another subset of these events associated with catastrophic (large and fast) landslides.

The event-detection algorithm locates events with an initial accuracy of 20 to 100 km. A terrestrial landslide source is established by combining this geographic location with satellite imagery, field photographs, news reports, local seismic recordings, and other sources. A comprehensive investigation of 195 unassociated detections for 2010 led to the identification of 11 major landslides (table S1, events 16 to 26). All of the seismically detected landslides generated long-period surface waves (SW) roughly equivalent to a magnitude $M_{SW} \sim 5$ tectonic earthquakes, and all were recorded at multiple seismographic stations. Tectonically generated surface-wave signals of this magnitude are routinely used to determine earthquake fault geometries and seismic moments, suggesting that similar methods could also be used to provide a quantitative characterization of the detected landslides. For example, Kanamori and co-workers measured a subhorizontal force of ~ 150 s duration and maximum amplitude $\sim 10^{13}$ N associated with the massive debris avalanche after the 1980 eruption of Mount St. Helens volcano (table S1). Seismological analyses of long-period data have usually focused on single landslide events and typically have been limited to estimation of the average slide direction (often only in the horizontal), peak force, and duration of sliding. Field observations, by contrast, frequently suggest complex three-dimensional (3D) landslide trajectories, and numerical modeling has highlighted the effects of such complexity on the radiated seismic waves.

We developed an inverse method to infer the 3D force sequence generated by bulk landslide motion—from which we can deduce the trajectory of slip and dynamic properties. The new algorithm builds on and extends established methods used in earthquake analysis. When applied to one of the largest landslides of 2010, this approach results in a first order characterization of the event (Fig. 17-1). On 4 January of that year, our algorithm automatically detected a seismic event of long-period magnitude $M_{SW} \approx 5.3$ at 08:36 GMT and roughly located the source in northern Pakistan (table S1). None of the international earthquake-monitoring agencies ISC, IDC, or NEIC reported this event. After anecdotal reports that a major landslide had struck the village of Attabad that morning—blocking the Karakoram Highway, damming the Hunza River, and causing several fatalities—we inspected long-period waveform data recorded on proximal stations and established that the seismic signal was likely caused by the Attabad slope failure. This association was confirmed by our inverse model, which provided a more accurate source location within 15 km of Attabad and which pointed to a direction of motion down to the south-southwest, consistent with local reports. These reports also indicated a time of failure consistent with the seismic detection.

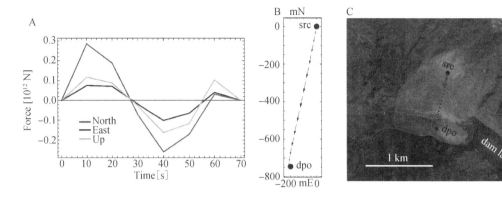

Fig. 17-1 Landslide force history and trajectory for the Hunza-Attabad landslide. (A) Inversion of the landslide force history $F(t)$ (LFH) of this event, pinning the time of main failure at 08:37 UT (table S1). (B) The planform trajectory of landslide motion deduced by doubly integrating the LFH and scaling by the runout distance mapped in (C). (C) Satellite-image mapping of the landslide scar and runout. The estimated centers of the source ("src") and deposits ("dpo") are indicated; their spatial separation was used to estimate Dh, determine the effective mass, and scale the displacement trajectory $D(t)$.

The estimated time sequence of forces induced by acceleration of the Hunza-Attabad landslide indicates a roughly sinusoidal sequence lasting $\Delta t \sim 60$ s (Fig. 17-1A). The 3D force vector components vary in a synchronous fashion, which suggests a consistent azimuth of acceleration and deceleration and therefore a linear runout. During the first 25 s the force vector points consistently to the north-northeast with an upward vertical component, indicating reaction to acceleration of the slide mass downhill in the south-southwest direction. The subsequent time series reflects reversal of the force vector during deceleration, as the slide mass approached the bottom of the valley.

(Cited from Ekström G, Stark C P. Simple scaling of catastrophic landslide dynamics [J]. Science, 2013, 339(6126): 1416-1419.)

New Words and Expressions

catastrophic	*adj.*	灾难的
debris	*n.*	碎片,碎屑
seismic	*adj.*	地震的
entrainment	*n.*	夹带,传输
magnitude	*n.*	震级
inverse	*n.*	相反,倒转
sinusoidal	*adj.*	正弦曲线的
vector	*n.*	矢量
landslide	*n.*	滑坡,山崩

tectonic	*adj*.	构造的
detected	*v*.	检测到的
volcanic	*adj*.	火山的
azimuth	*n*.	方位,方位角
trajectory	*n*.	轨迹,轨线
terrain	*n*.	地形,地势
glacier	*n*.	冰川

译文:灾难性滑坡动力学的简单比例尺度

滑坡地震辐射具有广谱性和复杂性。短周期波发生在粒状物质内并沿着其滑动边界发生。与构造地震中突然性应力下降相关的脉冲辐射相比,它们随时间分布且幅度低。滑坡辐射的长周期波更为简单:它们是由滑坡体大量加减速引起的固态土卸载和重新加载的宽幅循环而产生的。相应的动量交换因运动期间的传输和沉积以及沿滑动路径的地形起伏而变得复杂。在大型山体滑坡中,典型的卸载再加载时间为几十秒,这使得它们成为该量级周期的地震波的有效来源。

国家及国际机构进行的传统地震监测旨在探测脉冲型短周期地震波并确定相关构造地震和爆炸的位置,很少探测到山体滑坡事件。一种基于全球地震网(GSN)近实时数据的补充方法允许通过长周期波场的连续反投影来检测地震事件。该事件检测算法能检测到其他机构报告中90%以上的震级 $M \geqslant 5.0$ 的浅源地震,并且每月检测出约10个其他地震活动记录中没有的事件。在这些不相关的事件中,有些与大规模冰川裂冰及火山爆发有关。这里,我们鉴别并研究了与灾难性(大型和快速)滑坡相关的一类事件。

事件检测算法以20到100千米的初始精度来定位事件。通过将该地理位置与卫星图像、野外照片、新闻报道、当地地震记录以及其他来源记录相结合建立了陆地滑坡源。在对2010年的195次无关联检测进行了全面调查后,确定了11起主要滑坡(表S1,事件16至26)。所有由地震波探测到的滑坡产生的长周期表面波(SW)大致相当于一个 $M_{SW}\sim 5$ 级的构造地震,并且都被多个地震台站记录。这种震级的构造地震产生的表面波信号通常用于确定地震断层几何形状和地震发生时刻。这表明类似的方法也可用于对探测到的滑坡进行定量描述。例如,在1980年圣海伦火山喷发后,Kanamori及其同事测量出与大规模碎屑崩落有关的持续时间为～150 s,最大振幅为～10^{13} N 的近似水平力(表S1)。长周期数据的地震学分析通常集中在单个滑坡事件上,并且通常仅限于估计平均滑动方向(通常仅在水平方向上)、峰值和滑动持续时间。相比之下,野外观测常常显示出复杂的三维(3D)滑坡轨迹,而数值模拟强调了这种复杂性对辐射地震波的影响。

我们开发了一种反演法来推断由大块滑坡运动产生的三维力序列,并据此推导出滑动轨迹和动态特性。新算法是对地震分析中已有方法的继承和发展。当应用于2010年最大的一次山体滑坡时,这种方法能得到该事件的一阶特征(图17-1)。同年1月4日,我们的算法在格林尼治标准时间 08:36 时自动检测到长周期震级 $M_{SW}\approx5.3$ 的地震事件,大致位于巴基斯坦北部(表 S1)。而国际地震监测机构 ISC、IDC 或 NEIC 都没有报道过这一事件。据传闻,当天早晨阿塔巴德村发生了一次重大山体滑坡,阻塞了喀喇昆仑公路,拦截了罕萨河,造成数人死亡。我们检查了附近站点记录的长周期波形数据,并确定地震信号很可能是由阿塔巴德的滑坡引起的。这种关联被我们的反演模型证实了。该模型在村庄 15 公里范围内提供了更准确的震源位置,并指明了向南—西南方向的运动方向。这些信息与当地的报道是一致的。这些报告也表明了滑坡破坏时间与地震检测是一致的。

罕萨—阿塔巴德滑坡加速度引起的力的时程曲线大致为一个持续 60 秒的正弦波(图 17-1A)。三维力矢量的分量以同步方式变化,这表明加速和减速的方位角一致,因此是线性滑动。在前 25 秒期间,力矢量始终指向北—东北方向并具有向上的垂直分量,说明对滑坡体在南—西南方向的下坡加速度有反应。随后的时间序列反映了当滑坡体接近谷底时力矢量在减速期间的反转情况。

Text 18 Elements of Soil Physics

Tensile strength of water

Based on molecular considerations (hydrogen bonding), it has been concluded (Speedy, 1982) that tension in pure, deaired water cannot exceed 160 MPa; this value was confirmed experimentally by Zheng et al. (1991). Two points are particularly relevant with respect to this observation:

a) The tensile strength of 160 MPa is a result of homogeneous cavitation occurring in cases when the water does not contain cavitation nuclei. Homogeneous cavitation is usually identified with the tensile strength of pure, deaired water. This type of cavitation is obviously not relevant to unsaturated soils which are mixtures of solids, water, air and vapor, and therefore always contain cavitation nuclei.

b) As was pointed out above, the maximum theoretical and experimental value of negative potential in soil is of the order of 1,000 MPa, which is almost an order of magnitude greater than the tensile strength of pure water. It is clear, therefore, that the negative potential in soils cannot be identified with mechanical tensile stress in the soil water.

Skepticism with respect to the reality of high tensile stress in soil water is hardly new. Blight (1967) noted that the extremely large negative water potentials indicated by psychometric theory are incompatible with the tensile strength of water. He pointed out that "suction, as measured, must be considered merely as a convenient index of the affinity of the soil for free water" (rather than as a real mechanical tension). Koorevaar et al. (1983), stated that negative absolute water pressures (less than -100 kPa.), do not exist in soils. Iwata et al. (1995) were of the opinion that "soil water pressure below -1 bar does not have physical significance". For the present purpose, the following conclusion of Gray and Hassanizadeh (1991) is particularly relevant: "Although water, under certain conditions exhibits appreciable tensile strength, it does not do so in contact with air. When the (absolute) water pressure approaches the vapor pressure, the water will begin to evaporate." i.e. a cavitation process will be initiated, implying that the mechanical tension in the soil water is reduced to 80 to 100 kPa.

All experimental methods for the determination of "suction" in soil are based on the principle that, at equilibrium, the water potential in the soil and the measuring device are equal. Consequently suction, in the sense of tension in the soil water, is not a measurable quantity, and the actual measured quantity is water potential. Realizing this we use the terminology water potential when referring to experimental results in which the original term used was "suction", which implies water tension.

The soil-water characteristic curve (SWCC), is a relationship between water potential and the amount of water in the soil. This amount can be characterized by the gravimetric or volumetric water content (ω or θ) respectively, or by the degree of saturation, S. The SWCC is one of the main constitutive characteristic of unsaturated soils. Fig. 18-1 is a schematic soil-water characteristic curve presented by Lu and Likos (2004), (adopted from McQueen and Miller, 1974). Note that in this figure, the capillary regime (in which water can be in tension) extends from a suction of zero to 100 kPa. At lower potentials, the water is characterized as adsorbed. The potential in adsorbed water does not have stress characteristics; in fact it is doubtful if it is even possible to define the macroscopic variable, "pressure", in the thin adsorbed water films.

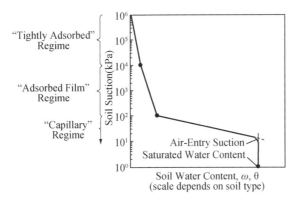

Fig. 18-1   Schematic physical characteristic of SWCC (after Lu and Likos, 2004,
reproduced by permission of John Wiley & Sons, Inc.)

The above review indicates that several authors in soil mechanics and physics (e.g. Koorevaar et al., 1983; Gray and Hassanizadeh, 1991; Iwata et al., 1995; Lu and Likos, 2004) suggested that water tension larger then 100 kPa does not exist in soils. Various authors justify their doubts on different bases. It seems reasonable to assume that heterogeneous and cavitation nuclei rich materials like unsaturated soils can cavitate under the water tension associated with the capillary potential. At the same time, it appears that the cavitation tension in fine grained soils may be higher than the classical value of 100 kPa, and it is not necessarily equal in all soils. These higher cavitation tensions in clays are probably the result of geometrical constraints imposed by the clay network, and various physical-chemical forces which become important only in materials having large specific surface areas. However, while the cavitation tension may be different in clays than in other materials, there is no doubt that it exists even in clays (e.g. Bishop et al., 1975). The exact value of the cavitation tension is not essential for the main physical arguments discussed here, although it will have an effect on the range of validity of the concepts presented.

Soil-water potentials

Buckingham (1907) introduced the notions of total, gravitational and capillary potentials as well as the representation of the SWCC. A formal, modern definition and discussion of the soil-water total potential density (i. e. energy per unit quantity of water), φ_T, at a point X in a porous medium is given by many authors (e.g. Edlefsen and Anderson, 1943; Bolt, 1976). This definition may be stated as: "φ_T at point X is the amount of work that must be done per unit quantity of pure water in order to transport, reversibly and isothermally, an infinitesimal quantity of water from a pool of pure water at a specific elevation and atmospheric pressure to the soil water at the point X under consideration".

Arbitrarily assigning the value $\varphi_T = 0$ to the reference pool, $\varphi_T(X)$ represents a potential energy density of soil water at point X. If the "unit quantity of water" in the above definition is taken as unit mass, φ_T is the chemical potential μ_T which has units of [J/kg]. Considering energy per unit weight of water, results in φ_T being expressed as the total head, h_T, which has units of length [m]. Reference to energy per unit volume of water yields a measure ψ_T, which has units of pressure [Pa]. It is noted, however, that the fact that ψ_T has pressure dimensions does not automatically make it a mechanical stress as implied by the geotechnical term, suction, (just as the head, h_T, is not a physical length). The relation between the three different forms of φ_T is $\psi_T = \rho_w \mu_T$, $\mu_T = g \rho_w h_T$ where: ρ_w is the mass density of water, and g is the acceleration of gravity. $\{\rho_w, g\}$ are constants, and the three different forms $\{\mu_T, \psi_T, h_T\}$ of φ_T are physically equivalent.

It is convenient to separate the total potential into a number of components:

$$\psi_T = \psi_m + \psi_{os} + \psi_{gr} + \cdots \tag{1}$$

where ψ_m is the matrix potential, representing the interaction of soil water with both the mineral soil skeleton and the gas phase; ψ_{os} is the osmotic potential which depends on the salt concentration in the bulk water; an d $\psi_{gr} = \rho g z$ is the gravitational potential, with z representing the elevation relative to an arbitrary reference datum. It is convenient to introduce the term internal potential $\psi_{int} = \psi_m + \psi_{os}$ which represents all potential components associated with sources which are internal to the soil element. Some devices for measuring water potential, (e.g. tensiometers), measure matrix potential, while others (e.g. psychrometers) measure internal potential. The only component not included in ψ_{int} is the gravitational potential ψ_{gr} which is not relevant to constitutive characterization of the soil. In this paper it is assumed that $|\psi_{os}| \ll |\psi_m|$, which is a reasonable assumption for many, non-saline soils.

(Cited from Baker R, Frydman S. Unsaturated soil mechanics: Critical review of physical foundations [J]. Engineering Geology, 2009, 106(1-2): 26-39.)

New Words and Expressions

deaired wate	*n.*	脱气水
cavitation	*n.*	空化、空穴作用
suction	*n.*	吸力
validity	*n.*	有效性
isothermally	*adv.*	等温地
mineral	*n.*	矿物
osmotic	*adj.*	渗透的
datum	*n.*	基点,基面
homogeneous	*adj.*	均匀的,同质的
incompatible	*adj.*	不相容的
evaporate	*v.*	蒸发
reversibly	*adv.*	可逆地
matrix	*n.*	基质
skeleton	*n.*	骨架
concentration	*n.*	溶液
non-saline	*adj.*	非盐渍的
adsorptive	*adj.*	吸附的

译文:土壤物理学的基本元素

水的抗拉强度

基于分子(氢键)考虑,纯脱气水的张力不会超过 160 MPa(Speedy,1982)。Zheng 等人(1991)通过实验证实了这个数值。关于这项观察,有两点特别重要:

a) 160 MPa 的抗拉强度是水在不含气核的情况下发生均匀空化的结果。均匀空化通常与纯脱气水的抗拉强度有关。这种空化类型显然与非饱和土无关,非饱和土是固体、水、空气和蒸汽的混合物,因此总是含有空化核。

b) 如上所述,土体中负电势最大的理论值和实验值约为 1 000 MPa,比纯水的抗拉强度几乎大一个数量级。因此,土体中的负电势很显然不能与土壤水的机械拉伸应力相等同。

对土壤水中存在高拉伸应力的事实持怀疑态度并不罕见。Blight(1967)指出,心理测量学理论显示极大的负水势与水的抗拉强度不相容。他指出,"正如测量的那样,吸力只被看作土对游离水亲和力的便利指标"(而不是真正的机械张力)。Koorevaar 等人(1983)表明土中不存在负绝对水压(低于 -100 kPa)。Iwata 等人(1995)认为"土壤水压低于 -1 bar 并没有物理意义"。就目前而言,Gray 和 Hassanizadeh(1991)的以下结论特别重要:"水尽管在某些条件下具有可观的抗拉强度,但在与空气接触时却不具备。当(绝对)水压接近蒸汽压时,水就开始蒸发。"即一个空化过程将会开始。这意味着土壤水中的机械张力会降低到 80 至 100 kPa。

用于测定土壤中"吸力"的所有实验方法都基于以下原理:在平衡时,土壤中的水势和测量装置的水势是相等的。因此,土壤水分中的张力是不可测量的量,实际测量的量是水势。认识到这一点,我们在引用实验结果时使用了术语水势代替原始术语"吸力",这就是水的张力。

土-水特征曲线(SWCC)是土中水势与含水量之间的关系。该量可以分别通过重量含水量或体积含水量(ω 或 θ)或饱和度 S 来表征。SWCC 是非饱和土的主要本构特征之一。图 18-1 是 Lu 和 Likos(2004)展示的土-水特征曲线示意图。注意,在该图中,毛细管状态(水可以处于拉伸状态)的吸力从 0 延伸到 100 kPa。在较低的水势下,水被表征为吸附水,吸附水的水势没有应力特征;事实上,在薄膜水中是否可以定义宏观变量"压力"是值得怀疑的。

上述评论表明,土力学和土壤物理学方面的几位作者(例如 Koorevaar et al.,1983;Gray and Hassanizadeh,1991;Iwata et al.,1995;Lu and Likos,2004)认为土中不存在水张力大于 100 kPa 的情况。不同的作者基于不同证据证明了他们的疑问。认为富含不均质气核的物质(如非饱和土)在与毛细水位势相关的水张力作用下会产生空穴的看法似乎是合理的。同时,细颗粒土中的空化张力似乎可能高于 100 kPa 的经典值,并且空化张力在各种土中并非相等。粘土中有较高的空化张力可能是由于粘土网状结构形成的几何约束,以及仅在具有大比表面积的物质中才变得重要的各种物理化学力的作用。虽然粘土中的空化张力可能与其他物质不同,但毫无疑问它存在于粘土中(例如 Bishop et al.,1975)。尽管空化张力的确切值会对所提出概念的有效范围产生影响,但它对于此处讨论的主要物理论点并不重要。

土-水势

Buckingham(1907)介绍了总势、重力势和毛细管势的概念以及 SWCC 的表示。许多作者对多孔介质中 X 点的土水总势能密度(即每单位水量的能量)φ_T 给出了正式和现代的定义及其相关讨论。该定义可以表述为:"X 点处的 φ_T 是每单位纯水为了在一定的高度和大气压条件下可逆且等温地运移到土壤水所在的 X 点而必须做的功"。

任意将某处取 $\varphi_T=0$,$\varphi_T(X)$ 表示在点 X 处土水势能密度。如果将上述定义中的"单位水量"作为单位质量,则 φ_T 是化学势 μ_T,单位为 [J/kg]。考虑到每单位重量水的能量,φ_T 表示为总水头 h_T,其具有长度单位 [m]。参考每单位体积水的能量产生测量值 ψ_T,其具有压力单位 [Pa]。然而,值得注意的是,ψ_T 具有压力尺度但不是土工术语所指的机械应力,吸力(正如水头 h_T 不是物理长度)。φ_T 的三种不同形式之间的关系是:$\psi_T=\rho_w\mu_T$,$\mu_T=g\rho_wh_T$。其中:ρ_w 是水的质量密度,g 是重力加速度。$\{\rho_w,g\}$ 是常数,φ_T 的三种不同形式 $\{\mu_T,\psi_T,h_T,\}$ 是等价的物理量。

我们可以方便地将总势能分成多个组成部分:

$$\psi_T=\psi_m+\psi_{os}+\psi_{gr}+\cdots \tag{1}$$

其中 ψ_m 是基质势,表示土壤水与土壤矿物骨架以及气相的相互作用;ψ_{os} 是渗透势,它取决于水中的盐浓度;$\psi_{gr}=\rho gz$ 是重力势,z 表示相对于任意参考基准的高程。这里引入内部势能 $\psi_{int}=\psi_m+\psi_{os}$,其表示土壤元素内部所有势能组成。有用于测量水势的装置(例如张力计)和测量基质势的装置,也有其他装置(例如湿度计)是用于测量内部势能的。在 ψ_{int} 中,唯一不包括的成分是重力势 ψ_{gr},它与土壤的本构特征无关。本文假设 $|\psi_{os}|\ll|\psi_m|$,这是对许多非盐渍土的合理假设。

Text 19　World Stress Map Project

In this chapter, a brief history of the World Stress Map (WSM) project is given, and the global database for contemporary tectonic stress orientation data of the Earth's crust is presented. This includes the statistics of stress data related to the stress determination method used, and the distribution of stress records versus depth. The quality ranking of WSM data most appropriate to rock mechanics and rock engineering purpose is presented and discussed.

Historical Aspects

In the WSM project, a global database for contemporary tectonic stress data of the

Earth's crust is compiled. It was originally compiled by a research group as part of the International Lithosphere Programme (Zoback et al., 1989). During the time period 1995—2008, the WSM Project was a research project of the Heidelberg Academy of Science and Humanities, Germany located at the Institute of Geophysics at Karlsruhe University, Germany (Heidbach et al., 2008). Since 2009, the home of the World Stress Map Project is the German Research Centre for Geosciences (GFZ) at Potsdam, Germany (Heidbach et al., 2010).

The first release of the WSM database contained 3,574 data records (Zoback et al., 1989) and the second one approximately 7,300 data records (Zoback, 1992). The major findings of the first phase of the WSM project are published in a special Volume of the Journal of Geophysical Research edited by Mary Lou Zoback (JGR Vol. 97). In this first phase of WSM, the compilation of data was primarily hypothesis-driven to investigate the plate boundary forces (including mantle drag) and to answer the question to what extend these forces are causing the long wave-length stress patterns. The focus of WSM was on stress data reflecting the large-scale contemporary tectonic intraplate or midplate stress field (i.e., plate scale stresses) rather than details and complexities close to plate boundaries where the overall kinematics and deformation is quite well known (Muller et al., 1992; Zoback, 1992). In the second phase of the WSM project between 1996 and 2008, this philosophy has changed to a data-driven compilation. In this period, there was an almost three-fold increase of stress data records; the current WSM database contains 21,750 data records, see Fig. 19-1. A few of the major conclusions derived from the analysis of the information in the WSM are that (1) over broad regions in the interior of many plates of the Earth's lithosphere are characterized by uniformly and consistently oriented horizontal stress fields like eastern North America, western Europe, the Andes, the Aegean; (2) most mid-plate or intra-plate continental regions are dominated by compressive stress regimes in which one or both horizontal stresses are greater than the vertical stress; and (3) in continental extensional stress regimes where normal or strike-slip stress field exist the maximum principal stress is vertical and generally occur in topographically high area. In the compilation shown in Fig. 19-1, all stress information of sufficient quality is added to the database regardless if the individual stress data point is representative for a larger region or not. In particular, the sedimentary basin initiative leads to a major increase of stress data records in areas that very much likely do not represent the long-wave length pattern, but local stress heterogeneities instead (Roth and Fleckenstein, 2001; Tingay et al., 2005a, b, 2009; Heidbach et al., 2007). This evolution of WSM database into more local and shallow details of the in situ stress makes the project more attractive for geoengineering and geotechnical applications (Fuchs and Muller, 2001; Tingay et al., 2005b; Henk, 2008).

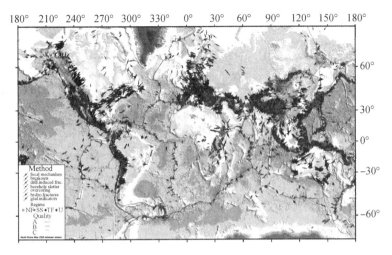

Fig. 19-1   **World Stress Map based on A—C quality data records of the WSM database release 2008, excluding all Possible plate Boundary Events (PBE) (Heidbach et al., 2010). Bars with symbol indicate SH orientations according to stress determination technique, and bar length is proportional to data quality Colors indicate stress regimes with red for normal faulting (NF), green for strike-slip faulting (SS), blue for thrust faulting (TF), and black for unknown regime (U). Plate boundaries are taken from the global model PB2002 of Bird (2003)**

Various academic and industrial institutions working in different disciplines of Earth sciences such as geodynamics, hydrocarbon exploitations and rock engineering use the World Stress Map. The uniformity and quality of the WSM data is guaranteed through (1) quality ranking of the data according to an internationally accepted scheme, (2) standardized regime assignment and (3) analysis guidelines for various stress indicator. To determine the SH orientation, different types of stress indicators are used in the WSM database. The 21,750 data records of the latest database release 2008 are grouped into four major categories with the following percentage values (Heidbach et al., 2010): (1) earthquake focal mechanisms (72%), (2) wellbore breakouts and drilling induced fractures (20%), (3) in situ stress measurements like overcoring, hydraulic fracturing and borehole slotter (4%), and (4) young geologic data from fault slip analysis and volcanic vent alignments (4%). In Fig. 19-2, A-C quality data are separated into the percentage values of each individual method used for stress estimation. In Fig. 19-2a, the dominating portion of the pie chart are stress data from earthquakes (73%), which show a depth distribution down to 40 km. In Fig. 19-2b, the focus is on borehole related stress data and geological stress indicator valid for shallow depth above 6 km [27% of the pie chart from Fig. 19-2(a)]. Individual stress indicators reflect the stress field of different rock volumes (Ljunggren et al., 2003), and different depths ranging from surface to 40 km depth (Heidbach et al., 2007). Fault plane solutions related to large earthquakes (Angelier, 2002) provide the majority of data. Below 6 km depth, earthquakes are the only stress indicators available, except from a few ultra deep drilling projects [Fig. 19-2(b)]. In general, the relatively small percentage of in situ stress measurements is due to the demanding quality of the ranking scheme and the fact that many of the data are company owned.

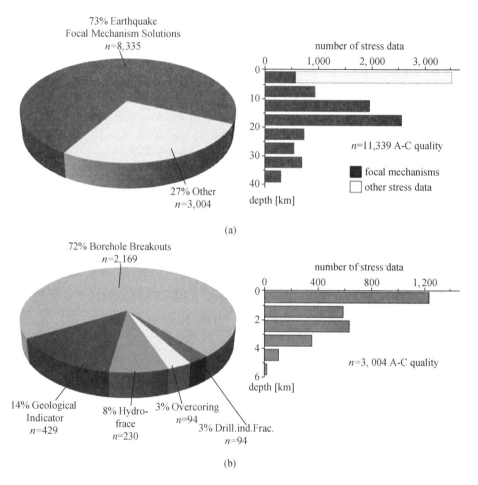

Fig. 19-2　Statistics of S_H orientation data with A-C quality ($n = 11,339$) data records. a With and b without earthquake related stress records ($n = 3,004$). *Left* pie charts, *right* depth sections

(Cited from Zang A, Stephansson O, Heidbach O, et al. World stress map database as a resource for rock mechanics and rock engineering [J]. Geotechnical and Geological Engineering, 2012, 30(3): 625-646.)

New Words and Expressions

database	*n.*	数据库
distribution	*n.*	分布
mantle	*n.*	地幔

interior	*n.*	内部
sedimentary	*adj.*	沉积的
uniformity	*n.*	一致性
borehole	*n.*	钻孔
crust	*n.*	地壳
originally	*adv.*	最初
complexities	*n.*	复杂性
topographically	*adv.*	从地形上
disciplines	*n.*	学科
indicator	*n.*	指示器,标志
shallow	*n.*	浅层,浅部

译文:世界应力图项目

本章介绍了世界应力图(WSM)项目的简要历史,并展示了包含地壳现代构造应力方向数据的全球数据库。这其中包括与所用应力确定方法相关的应力数据的统计信息,以及应力记录沿深度分布的信息,同时提出并讨论了最适合岩石力学和岩石工程目的的世界应力图的数据质量等级的划分方法。

历史方面

在 WSM 项目中,编制了一个关于地壳现代构造应力数据的全球数据库。作为国际岩石圈计划的一部分,它最初由一个研究小组编制(Zoback et al.,1989)。在 1995~2008 年期间,WSM 项目是德国卡尔斯鲁厄大学地球物理研究所海德堡科学与人文科学院的一个研究项目(Heidbach et al.,2008)。自 2009 年以来,WSM 项目的所在地则是德国波茨坦的德国地球科学研究中心(GFZ)(Heidbach et al.,2010)。

WSM 数据库的第一个版本含 3 574 条数据记录(Zoback et al.,1989),第二个版本包含大约 7 300 条数据记录(Zoback,1992)。该项目初始阶段的主要研究成果发表在由 **Mary Lou Zoback** 编辑的地球物理研究期刊上(JGR Vol. 97)。在 WSM 的初始阶段,数据汇编主要通过假设驱动来研究板块边界力(包括地幔拖曳),并以此解释这些力在多大程度上导致了长波长应力模式的问题。WSM 的研究重点是反映大规模现代构造板块内或板块中应力场的应力数据(即板块应力),而不是众所周知的靠近板块边界处的复杂整体运动和变形细节(Muller et al.,1992;Zoback,1992)。在 1996 年至 2008 年 WSM 项目的第二阶段,这种理念已经转变为数据驱动的编译。在此期间,应力数据记录量增加了近三倍;当前的 WSM 数据库包含了 21 750 条数据记录,见图 19-1。从 WSM 数据中分析得出的一些主要结论是:(1)地球岩石圈中许多板块内部的广泛区域存在均匀一致的水平应力场,如北美东部、西部欧洲、安第斯山脉、爱琴海;(2)大多数中部板块或者内陆区域以压应力状态为主,其中一个或两个水平方向应力大于垂直方向应力;(3)在正断层或走滑断层应力场所赋存的大陆拉伸应力状态下,最大主应力是垂直的,通常出现在地形高的区域。在图 19-1 中,无论单个应力数据点是否能代表更大的区域,所有具有足够高质量的应力信息都被添加到数据库中。特别是沉积盆地导致应力数据记录大幅增加,这些区域很有可能不代表长波长模式,而是局部应力异质性(Roth and Fleckenstein,2001;Tingay et al.,2005a,b,2009;Heidbach et al.,2007)。WSM 数据库向更局部、更浅层的地应力信息演变,使该项目对地球工程和岩土工程应用更具吸引力(Fuchs and Muller,2001;Tingay et al.,2005b;Henk,2008)。

从事如地球动力学、碳氢化合物开发和岩石工程等不同地球科学学科的各种学术和工业机构,都在使用世界应力图。WSM 数据的一致性和高质量主要通过以下三点来保障:(1)根据国际公认的方案对数据进行质量分级;(2)标准化的制度安排;(3)各种应力指标的分析指南。为了确定应力方向,WSM 数据库中使用了不同类型的应力指标。2008 年最新数据库发布的 21 750 条数据记录分为四大类,其百分比值如下(Heidbach et al.,2010):(1)地震震源机制(72%);(2)井孔爆破和钻井诱发裂缝(20%);(3)地层应力测量,如覆岩,水力压裂和钻孔开槽(4%);(4)断层滑动分析和火山口较新的地质数据(4%)。在图 19-2 中,A-C 质量数据被分成用于应力估计的每种单独方法的百分比值。在图 19-2(a)中,饼图的主要部分来自地震的应力数据(73%),其分布深度达到 40 千米。在图 19-2(b)中,重点是钻孔相关的应力数据和地质应力指标,且对 6 千米以上的浅层深度有效[图 19-2(a)中饼图的 27%]。单个应力指标反映了不同岩石体积的应力场(Ljunggren et al.,2003),以及从地表延伸到 40 千米深范围内的不同深度(Heidbach et al.,2007)。与大地震有关的断层面解决方案(Angelier,2002)提供了大部分数据。深度低于 6 千米时,除了一些超深钻井项目,地震是唯一可用的应力指标[图 19-2(b)]。一般而言,现场地应力测量数据的百分比相对较小是因为分级方案对质量的要求以及许多数据归公司所有。

Text 20　Can Buildings Be Made Earthquake-Safe?

The 100th anniversary of the 1906 San Franciso earthquake provides an opportunity to reflect on what we have learned about earthquake-resistant design and how research in the field can be used to build better in the future. During a major earthquake, ground shaking under an urbanized area can cause serious damage or even the collapse of buildings and bridges, freeways, power lines, and other critical infrastructure. The damage can lead to a disaster threatening thousands of lives, and affecting the economic well-being of the epicentral region. Over the past 100 years, architects and engineers have developed tools to assess the earthquake hazard, as well as the stability of individual designs. Building codes provide the key policy mechanism for regulating a standard of safety. However, the death toll in recent earthquakes in Pakistan, Iran, Turkey, and India, and the economic losses in Japan and the United States, suggest that the technical capacity to design buildings to withstand earthquake forces may not be all that is needed to make society safe in earthquakes. Throughout history, designers and builders have attempted to learn from earthquake damage in order to improve the stability of their structures. The Roman scientist, Pliny the Elder, suggested digging underground to release the earth's trapped air. When Lima was leveled by an earthquake in 1746, authorities limited the height of buildings and required quincha (adobe reinforced with bamboo) construction. In Portugal, the gaiola, a masonry building reinforced with a wooden framework, was perfected after the 1755 Lisbon earthquake. Between 1850 and 1906, architects and civil engineers in San Francisco attempted a variety of design innovations to reinforce their buildings for seismic stability. In one case, a raft of redwood logs was used to "float" a foundation on bay mud; other inventors patented "earthquake-proof" technologies. Others used iron tie-rods secured to the masonry walls to ensure that the exterior walls would not fail. In fact, photos of the destruction wrought by the 1906 earthquake and fire also show that numerous major brick buildings survived the shaking intact. All were designed by professionals who were extremely cognizant of the possibility of damage from lateral forces.

Still, the assessment of damage after the 1906 earthquake and fire forced architects and engineers to reassess the dangers and limits of unreinforced masonry (brick, stone, or adobe) construction, and to look to concrete and steel—new materials with the capacity to provide larger interior spaces, large openings for better light, as well as more efficient and safer construction for urban buildings. At the same time that the modern movement in architecture created the thin column and light slab for aesthetic and functional simplicity, engineers developed an understanding of the site effects and ground failures caused by earthquakes; and structural engineering began to understand the relation between strong motion, building configuration, and building performance. If it is possible to summarize the achievements of 20th-century earthquake engineering in a sentence, it would be to say that intuitive understandings of building behavior were replaced with scientific tools to measure ground shaking and mathematical models to analyze and predict structural behavior.

These included instruments designed to measure strong motion, laboratory shaking tables designed to simulate and test the response of building components to strong shaking, and mathematical (and later digital) models for calculating the strength of materials, building components, and joints. The development of the tools to collect data on the hazard and the use of analytic procedures to predict structural behavior under earthquake loads, combined with systematic investigations of building failures after events, led to an understanding of how to build both "strength" and "toughness" into buildings and other structures. Strengthening often requires making a building stiff, but stiffness can transfer the load to other building systems, causing extensive damage to partitions, mechanical systems, or contents. As such, design for seismic forces requires balancing the probable ground motions at the site and soil conditions with building configuration, height, materials, and structural system to meet a code standard or a client's performance expectations.

A downtown office building built by a developer will be designed per the code to provide safety for its occupants, and the owner accepts that certain systems may sustain damage and require time for repairs after an earthquake. However, an art museum, or a high-technology manufacturing facility, may require a design that protects the contents from damage and provides continued operations after an event. To meet such performance needs, new technologies have been developed in recent years. For example, base-isolation uses rubber or neoprene bearings to decouple the ground motion from the building structure. Dampers can be used in a structural frame to act as shock absorbers—again, to lessen the transmission of forces and limit damage.

Public-policy mechanisms are also a key component in limiting earthquake damage. Although building codes promote "safe development" in general, special requirements for schools, hospitals, and emergency-service facilities acknowledge the need for higher standards in certain public buildings. Land-use policies also reduce risk by restricting development in fault zones or on specific soil types. Similarly, government programs aimed at preparedness and hazards mitigation contribute to the reduction of losses.

Performance-based earthquake engineering is a relatively new concept that challenges engineers to rethink the parameters by which they measure good seismic design, and it challenges the policy-makers to rethink safety standards. The goal of performance-based earthquake engineering is to design facilities with predictable levels of seismic performance, using casualties, cost, and downtime as metrics. Current research at the Pacific Earthquake Engineering Research (PEER) Center combines a probabilistic assessment of the hazard, using suites of ground motions, with nonlinear dynamic analyses of building responses to determine a range of possible outcomes. Although seemingly straightforward, the methodology requires substantial data to adequately model the performance of all building components, and it requires critical judgment in quantifying uncertainty. Equally important, performance-based earthquake engineering requires consumers to specify clear performance goals, and local and state governments to adopt very different approaches to regulating construction. Performance-based earthquake engineering is both inspiring and daunting, but it will unquestionably change the way we design and build in the next 50 years.

The influence of computational technologies on design cannot be underestimated. Not only can designers analyze the technical performance of structures, but they can also use computational technologies to better design the day-lighting and thermal comfort zones in buildings, and to manage maintenance and operations. Computational tools are also changing the form of structures. The new technologies do not require space to be rectilinear for computational purposes. Digital design means that complex curved spaces can be represented, analyzed, and built to the highest seismic standards. The new de Young fine arts museum in San Francisco, and the Walt Disney Concert Hall in Los Angeles, are only two examples.

(Cited from Comerio M C. Can buildings be made earthquake-safe? [J]. Science, 2006, 312(5771): 204-206.)

New Words and Expressions

earthquake-resistant design		抗震设计
infrastructure	*n.*	基础设施
epicentral	*adj.*	震中的
quincha	*n.*	一种传统的建筑系统,基本上使用木材和甘蔗或巨型芦苇,形成一个用泥土和石膏覆盖的抗震框架
masonry	*n.*	砖石建筑
bay mud		海湾泥
lateral	*adj.*	横向的,侧面的
neoprene	*n.*	氯丁(二烯)橡胶
damper	*n.*	减震器
transmission	*n.*	传输
mitigation	*n.*	缓解
performance-based earthquake engineering		基于性能的地震工程
metric	*n.*	衡量标准
daunting	*adj.*	令人畏惧的
rectilinear	*adj.*	线性的

译文:建筑物能安全抗震吗?

1906 年旧金山地震的 100 周年纪念活动为我们提供了一个契机来反思我们对抗震设计

的了解,以及未来该如何把抗震领域的研究成果更好地应用于工程建设。在一场大地震中,城市地区的地面晃动会造成严重的破坏,甚至引起房屋、桥梁、高速公路、电线和其他重要基础设施的倒塌。这种破坏可能导致一场灾难,不仅威胁着成千上万人的生命,还影响着震中区的经济福祉。在过去的 100 年里,建筑工程师们研发了评估地震灾害和建筑设计稳定性的工具。建筑规范为控制安全标准提供了关键的政策机制。然而,最近发生在巴基斯坦、伊朗、土耳其和印度的地震造成的死亡人数以及发生在日本和美国的地震造成的经济损失表明,设计抗震建筑的技术能力可能并不是保持地震中城市安全所需的全部。纵观历史,设计师和建筑师都试图从地震破坏中吸取教训,以提高结构的稳定性。罗马科学家 Pliny the Elder 建议在地下挖掘,以释放地下残存的气体。当利马在 1746 年被地震夷为平地时,当局限制了建筑物的高度,并要求使用 quincha(即:竹加筋土坯)的建筑形式。在葡萄牙的加约拉,一个用木框架加固的砖石建筑在 1755 年里斯本地震后得到完善。在 1850 年至 1906 年间,旧金山的建筑师和工程师曾试图进行各种设计创新,加强建筑物在地震中的稳定性。在一个案例中,一堆红木被用来在海湾泥浆上"漂浮"一个基础;其他发明家申请了"抗震"技术专利。另一些人将铁杆固定在砖墙上,确保外墙不会倒塌。事实上,1906 年地震和火灾造成的破坏照片也显示,许多主要的砖房在地震中完好无损。所有这些(完好无损的砖房)都是由专业人员设计的,他们非常清楚侧向力可能造成的损害。

不过,1906 年地震和火灾后的损伤评估促使建筑师和工程师们重新评估未加固砌体(砖、石或土坯)建筑的危险和极限,并寻求利用混凝土和钢材等新型材料进行建设。这些新材料能够为城市建筑提供更大的内部空间,获得更充足光线的大开口以及更高效安全的构造。与此同时,现代建筑运动创造了具有美感且功能简单的薄柱和轻板,工程师们对场地效应和地震造成的地基破坏有了进一步认识;针对结构工程,工程师们也开始了解强震、建筑结构和建筑性能之间的关系。如果有可能用一句话来概括 20 世纪地震工程的成就,那就是用科学工具和数学模型来取代对建筑行为的直观理解,来测量地面震动和分析预测结构行为。

这其中包括测量强震的仪器、模拟和测试建筑构件对强震动力响应的室内振动台以及计算材料、建筑构件和接头强度的数学(和后来的数字化)模型。灾害数据收集工具的研发,能够进行地震荷载作用下结构行为预测的分析程序,再加上对震后建筑物破坏的系统调查,使人们懂得如何建造兼具"强度"和"韧性"的建筑以及其他构筑物。加固(建筑物)往往需要使建筑物变得坚硬,但刚度会将负载转移到其他建筑系统,对隔墙、机械系统或系统内部造成大面积损坏。因此,地震力的设计需要在现场可能出现的地面震动、土质条件与建筑物的布局、高度、材料和结构系统之间取得平衡,从而满足规范标准或客户的性能预期。

开发商建造的市区办公楼将按照规范进行设计,为住户提供安全保障。业主也接受当地震发生后,某些系统可能受到损坏,并需要时间进行维修。然而,艺术博物馆或高科技制造设施可能需要在地震发生后保护(设施)内部免受损害并能持续运作的设计。为了满足这些性能需求,近年来开发了新技术。例如,基础隔震法使用橡胶或氯丁橡胶轴承将地面震动与建筑结构分离。阻尼器可以在结构框架中充当减震器——同样,可以减少力的传递并减小损伤。

　　公共政策机制也是减小地震破坏的一个关键组成部分。虽然建筑法规总体上促进了"安全发展",但是对于学校、医院和紧急服务设施等某些公共建筑来说,由于对它们有特殊要求,因此需要更高的抗震标准。土地使用政策还通过限制在断裂带区域或特殊土壤类型上进行开发来降低风险。同样,旨在预防和减轻危害的政府方案也有助于减少损失。

　　基于性能的地震工程是一个相对较新的概念。它要求工程师重新思考用来衡量良好抗震设计的参数,并要求决策者重新思考安全标准。基于性能的地震工程的目标是设计具有可预测的抗震性能水平的设施,并以人员伤亡、经济损失和停工时间为衡量标准。太平洋地震工程研究中心目前的研究结合对灾害的概率评估,使用成组的地面震动,对建筑物响应进行非线性动态分析来确定一系列可能的结果。该方法虽然看似简单,但需要大量数据来充分模拟所有建筑构件的性能,并且需要在量化不确定性方面做出关键判断。同样重要的是,基于性能的地震工程要求客户明确性能目标,也要求地方和州政府采用截然不同的方式来规范建筑。基于性能的地震工程既鼓舞人心,又令人生畏,但它无疑将改变我们未来50年的设计和施工方式。

　　计算机技术对设计的影响不可低估。设计人员不仅可以分析结构的技术性能,还可以使用计算技术来更好地设计建筑物中的日光照明和热舒适区,并管理日常维护和运营。计算工具也在改变结构形式。新技术不需要为了避免复杂的计算而只使用直线型空间。数字设计意味着复杂的曲面空间可以按照最高的抗震标准呈现、分析和施工。旧金山新德扬美术博物馆和洛杉矶的华特迪士尼音乐厅就是两个例子。

Unit 5 Engineering Geology

Text 21 Assessing Landslide Hazards

On 31 May 1970, a large earthquake shook the highest part of the Peruvian Andes. Millions of cubic meters of rock dislodged from a mountainside and initiated a rock avalanche that traveled more than 14 km in 3 min, burying a city and killing more than 25,000 people. On 17 February 2006, a landslide of 15 million m³ that initiated on a slope weakened by long-term tectonic activity buried more than 1,100 people on Leyte Island in the Philippines.

Landslides such as these are a hazard in almost all countries, causing billions of dollars of damage and many casualties. Landslides also contribute to landscape evolution and erosion in mountainous regions (see the first figure). Here we discuss the latest

Fig. 21 - 1 Landslides in mountain regions. These rock avalanches were triggered by the 1980 Mammoth Lakes earthquake sequence. Several thousand rock falls and slides were associated with this event in central California.

strategies used to assess and mitigate landslide hazards.

The basic physics governing the initiation of landslides—the interactions among material strength, gravitational stress, external forces, and pore-fluid pressure—has been well understood for decades. The factors that govern whether landslide movements, once begun, will be catastrophic are less well understood. Nonetheless, much recent progress has been made in understanding those factors, as exemplified by basic research on fracture development in brittle materials and on the properties of flowing material.

Major causes of landslides are also well known, and these include rainfall, seismic shaking, human construction activities, landscape alteration, and natural processes of erosion that undermine slopes. Yet predicting just where and when a landslide will occur continues to be a complex proposition, because the properties of earth materials and slope conditions vary greatly over short distances, and the timing, location, and intensity of triggering events—such as storm precipitation or earthquake shaking—are difficult to forecast.

Two landslides at La Conchita in California illustrate the complexity of landslide occurrence and behavior. In 1995, a landslide consisting of a relatively coherent block of earth at La Conchita caused property damage but no fatalities. Ten years later, another landslide remobilized from the 1995 deposit, transformed rapidly into a highly fluid debris flow, and traveled downslope at a speed of 5 to 10 m/s, causing 10 fatalities (see the second figure).

Fig. 21-2   Dangerous complexity. This landslide at La Conchita, California, on 10 January 2005 destroyed 13 houses, severely damaged 23 others, and killed 10 people.

Current landslide hazard analyses and mitigation strategies tend to concentrate at one of two scales: intensive, site-specific analyses of individual slopes or landslide bodies, and regional-scale evaluations that seek to identify hazardous zones that are best avoided when construction is planned.

In a site-specific landslide evaluation, instruments may be installed into the slope to determine water pressures, measure subsurface slippage, and monitor surface deformation. Materials may be sampled for laboratory testing of shear strengths and other properties such as mineralogy and density. Because these methods are expensive, extensive and site-specific analyses are commonly restricted to slopes where the costs of construction, potential for damage, or risk to population justify the expense.

A range of analytical techniques is used to evaluate the potential for landslide initiation at the site-specific scale. The decades-old and generalized limit-equilibrium method envisions a landslide as a rigid sliding block, and this has proved useful for many engineering and construction applications. Some newer, more sophisticated methods are specialized for the analysis of such processes as volcano flank collapses and initiation of debris flows. In the case of volcano-flank collapses, for example, these new methods incorporate coupled numerical modeling of heat and groundwater flow to analyze the potential for landslide initiation involving steep volcano flanks due to hydrothermal pressurization. Such modeling predicts the occurrence of deep-seated landslides that match the dimensions of many observed landslides, whereas more traditional slope-stability analyses predict that the landslides would be shallow.

Regional-scale evaluations of landslide hazards also use a range of analytical techniques. For example, modeling that combines analysis of groundwater flow with slope stability calculations has been used to predict the timing and location of shallow, precipitation triggered landslides, and the Newmark analysis (which combines slope stability calculations with seismic ground-motion records) is widely used to evaluate the potential for landslides that could be triggered by earthquake shaking. Regional-scale analyses may also include empirical methods based on mapping landslide occurrences and developing statistical correlations among landslide occurrence, material and slope properties (such as rock type and slope steepness), and the strength of triggering events such as seismic shaking or rainfall intensity and duration.

Regional-scale landslide analyses took a leap forward with the advent of high-resolution remote-sensing imagery and the use of geographic information systems (GIS) technology. The first automated event-based mapping of landslides from satellite imagery was carried out after the 1999 Chi-Chi earthquake in Taiwan. More recently, landslides triggered by the 2004 Niigata Ken Chuetsu earthquake in Japan were mapped using a similar technique. Further automated landslide mapping of this kind would greatly extend the database on which regional-scale hazard and risk models may be constructed.

Several other techniques also have promise for increasing the accuracy, precision, and effectiveness of landslide hazard evaluation. For example, synthetic aperture radar interferometry can be used for early detection of landslide movements. Models are being developed to predict landslide motion based on detailed analyses of motion-induced changes in pore-fluid pressures and material properties in landslide shear zones. Finally, landslide warning systems can be used to issue public alerts and warnings for a particular region when accumulated and/or forecast amounts of rainfall equal or approach those amounts that have triggered landslides there in the past.

Current landslide research efforts around the world are generally small relative to the costs of landslide damage. A recent report by the U.S. National Research Council recommended a 15-fold increase in funding for landslide research and development in the United States. Although landslide hazard evaluation and mitigation strategies are advancing in many fundamental areas, the loss of life and destruction of property by landslides around the world will probably continue to rise as the world population increases, urban areas of many large cities impinge more on steep slopes, and deforestation and other human landscape alterations affect ever-larger areas.

(Cited from Keefer D K, Larsen M C. Assessing landslide hazards [J]. Science, 2007: 1136-1138.)

New Words and Expressions

dislodge from		从……中脱离
avalanche	v.	崩塌
tectonic	adj.	构造的
erosion	n.	侵蚀
catastrophic	adj.	灾难性的
fracture	n.	断裂
brittle	adj.	易碎的
precipitation	n.	降水量
coherent	adj.	连贯的,一致的
deposit	n.	沉积物
slippage	n.	滑动量
mineralogy	n.	矿物学
limit-equilibrium		极限平衡
sophisticated	adj.	复杂的
flank	n.	侧边

debris flow		泥石流
impinge	v.	对……起作用,影响
deforestation	n.	森林砍伐

译文:滑坡危害性评估

1970 年 5 月 31 日,在秘鲁安第斯山脉的最高处发生了大地震。数百万立方米的岩石从山坡上坠落,引发了山崩,在 3 分钟内滑动了超过 14 公里,掩埋了一座城市,造成 25 000 多人死亡。2006 年 2 月 17 日,在被长期构造活动削弱的斜坡上发生了 1 500 万立方米的山体滑坡,导致菲律宾莱特岛上的 1 100 多人被滑坡体掩埋。

这样的山体滑坡在几乎所有国家都是一种危险的灾害,且会造成数十亿美元的经济损失和大量的人员伤亡。山体滑坡也会促进山区地貌的演化和侵蚀(见第一幅图)。在这里,我们将讨论评估和减轻山体滑坡灾害的最新策略。

控制滑坡发生的基本物理机制——物质强度、重力、外力和孔隙水压力之间的相互作用——几十年来一直很清楚。但是滑坡运动一旦启动后,是否会引发重大灾难的控制因素就不太清楚了。然而,最近研究人员在对这些因素的理解方面取得了大量的进展,例如在脆性材料的断裂发育和流动材料的特性方面取得的基础性研究进展。

引发山体滑坡的主要原因也是众所周知的,包括降雨、地震、人类建筑活动、地形地貌改造以及边坡破坏的自然侵蚀过程等。然而,预测山体滑坡发生的时间和地点仍然是一个复杂的问题,因为岩土材料和边坡条件的特性在很小的范围内就会有很大的差异。另外,引起滑坡的因素,如暴雨或地震,在时间、位置和强度方面也很难预测。

在加州 La Conchita 发生的两次滑坡显示了山体滑坡发生和行为的复杂性。1995 年,在 La Conchita 发生了由一个较为连续的土块构成的山体滑坡,造成了财产损失,但没有人员伤亡。十年后,在 1995 年的滑坡堆积物上又重新触发了一次山体滑坡,并迅速转变为高流动性的泥石流,以每秒 5 m 至 10 m 的速度向下游流动,造成了 10 人死亡(见第二幅图)。

目前滑坡灾害的分析和减灾策略往往集中在以下两个尺度之一:一是对个别斜坡或滑坡体进行密集的、特定地点的分析,二是进行区域规模的评估,确定在施工规划时最好避开的危险区域。

在特定场地的滑坡(安全)评估中,可以在边坡上安装仪器来测定水压,测量地表以下土体的滑移量,并监测地表变形量。还可以对材料进行取样,以便在实验室测试其剪切强度和其他性质,如矿物属性和密度。因为这些方法成本高昂,广泛的实地分析通常仅局限于那些建筑成本高,破坏可能性大或人员伤害风险高的斜坡。

有一系列分析技术被用于评估在特定地点范围内发生滑坡的可能性。几十年来广泛使用的广义极限平衡法假定滑坡体为刚性滑块。事实证明,这种方法对许多工程与施工应用都很有效。一些更新和更复杂的方法被专门用于分析火山的侧面塌陷和泥石流的启动等过程。例如,在火山侧面塌陷的情况下,这些新方法结合了热量和地下水渗流的耦合数值模拟,来分析由水热增压引起的陡峭火山侧面发生滑坡的可能性。这种模型预测到深层滑坡的发生,这与许多观测到的滑坡尺寸相吻合,而传统的边坡稳定性分析的预测结果则是应发生浅层滑坡。

滑坡灾害的区域尺度评估也使用了一系列分析技术。例如,将地下水渗流分析与边坡稳定性计算相结合的建模方法被用来预测由降水引发的浅层滑坡的发生时间和位置,而Newmark 分析方法(将边坡稳定性计算与地震地面运动记录相结合)被广泛用于评估地震引发滑坡发生的可能性。区域尺度分析还采用了一些经验方法。这些方法基于滑坡点的分布图建立了滑坡发生点、滑坡体材料和边坡的特性(如岩石类型和坡度)、触发事件的强度(如地震动或降雨的强度和持续时间)之间的统计相关性。

随着高分辨率遥感图像技术和地理信息系统(GIS)技术的应用,区域尺度的滑坡分析取得了飞跃。1999 年台湾集集地震后,人们首次利用卫星影像自动绘制山体滑坡的图像。最近,日本也采用类似的技术绘制了2004 年新泻县中越地震引发的山体滑坡的图像。滑坡图像自动化绘制技术的进一步发展将大大扩充滑坡灾害数据库,而在该数据库上可以构建区域尺度的灾害和风险模型。

其他几种技术也有望提高滑坡灾害评价的准确性、精确性和有效性。例如,合成孔径雷达干涉技术可用于滑坡运动的早期探测。在对滑坡剪切带内孔隙流体压力和材料性质的运动变化进行详细分析的基础上,人们建立了滑坡运动的预测模型。最终,当某一地区累积及/或预测的雨量与过去曾在该地区引发滑坡的雨量相等或接近时,滑坡预警系统能够向公众发出警报和警告。

目前在世界各地进行的滑坡研究工作与滑坡灾害造成的损失相比,总体而言规模较小。美国国家研究委员会最近的一份报告建议将美国山体滑坡的研究和发展资金增加15 倍。尽管滑坡灾害的评估和减灾战略正在许多基础领域推进,但是世界各地由山体滑坡造成的生命和财产损失仍然有可能继续增加。这是因为随着世界人口的增加,许多大城市的城区会越来越多地向陡峭的山坡发展,而森林砍伐以及人类其他活动对地形地貌的改变也会对越来越大的区域造成不利影响。

Text 22   Engineering Geological Assessment of the Obruk Dam Site

Introduction

The principal factors that constrain the geomechanical properties of rocks are geological structure, mineralogical composition, discontinuities, and degree of weathering. Accordingly, the geological and geotechnical properties of rocks that comprise the basement to major engineering structures (such as dams) should be determined in the field and in the laboratory prior to construction. Axis locations, the excavatibility of rocks underlying the projected traces of spillways, diversion tunnels and power tunnels, grouting conditions and the determination of support systems for tunnels are also significant from the standpoint of optimal project design. Engineering geological investigations and rock mechanics studies mainly include discontinuity surveying, core drilling in-situ testing (Özsan and Akın, 2002).

In the present study, the basalts that represent the basement to the Obruk dam—constructed on the Kızılırmak River approximately 50 km NW of Çorum in central Anatolia—were investigated from an engineering geological perspective (Fig. 22-1). This dam, which is of the earth-fill type and intended for energy production and irrigation, is situated 67 m above the thalweg, with a height of 126 m from base to top and a 12,000 hm³ body volume, and was planned to have 203 MW of power with an annual energy production of 473 GWh. In 1979 and 1983, within the framework of the Lower Kızılırmak Project, General Directorate of the State Hydraulic Works (DSI) prepared engineering geological planning reports on the basis of exploratory drilling in the area of the initial axis of the Obruk dam (DSI, 1979, 1983). Kılıç (1999) studied the geomechanical properties of the basalts in the area of the planned initial axis of the Obruk dam, and divided these rocks into five groups on the basis of degree of weathering. Koçbay (2003), through study of the characteristics and degree of basalt weathering in the Osmancık—Çorum area, proposed a classification related to decomposition. Further, Koçbay and Kılıç (2003) investigated the possible use of these basalts as a natural building material.

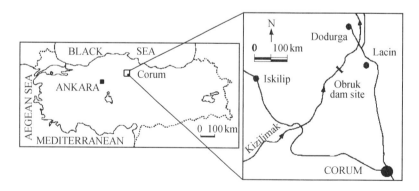

Fig. 22-1 Location map of the study area.

Because there is an active landslide located approximately 500 m on the source side of the initial axis area for the Obruk dam, it was relocated. Thus, the present study deals — from an engineering geological perspective — with the second axis area chosen. In order to study vertical and lateral variations in the basalts and take core samples thereof, do Lugeon tests in order to determine the coefficient of hydraulic conductivities, and to determine the ground-water level, we benefited from 17 basement investigation drill holes that were drilled by the General Directorate of the State Hydraulic Works (DSI) at construction sites for the dam project. These 17 drill holes had depths between 40 and 160 m and a total depth of 1,512 m. Furthermore, the predominant strike and dip directions of joints in the basalts were determined, and representative block samples were taken from outcrops.

After initially determining the mineralogical, petrographic and chemical characteristics of the samples, the samples were grouped on the basis of degree of weathering. Subsequently, the physical and geomechanical characteristics of 172 samples were ascertained. Using these data, an engineering geological map and sections for the traces of 1) the new axis location for the Obruk dam and 2) power and diversion tunnels were prepared. On the basis of all of these characteristics, the rock mass was classified from the perspective of tunneling.

Geology of the dam site

The Lutetian Bayat formation, comprising volcanites and flysch interbeds, crops out in the study area (MTA, 1975). In the vicinity of the dam site, the Bayat formation is widespread, and the predominant lithology is basalt. The basalt consists of a fine-grained dark matrix, medium-sized plagioclase crystals, relatively coarse-grained pyroxene and olivine phenocrysts, and opaque minerals. Locally, pyrite is present within the opaque phases. Chloritization is widespread in glassy rock flour and in pyroxenes, while plagioclases have been altered to clay. Carbonatization, silicification, chloritization and argillization increases in relation to degree of hydrothermal alteration. Alteration is more intense along discontinuities, and gypsum fillings occur in fractures. At the top of the sequence is Quaternary alluvium made up of block, gravel, sand, silt and clay-sized materials (Fig. 22-2).

The study area — located approximately 35 km from the North Anatolian Fault Zone (NAFZ), one of the most important tectonic features of Turkey — has been strongly affected by tectonism. Numerous NE-SW- and NW-SE-trending normal faults at varying scales occur to the right and left of the dam axis. The effects of tectonism and the development of joints are obvious. An engineering geological map of the dam site, diversion and power tunnels, spillway area, and right and left slopes, and sections thereof, are given in Figs. 22-2 and 22-3, respectively.

105

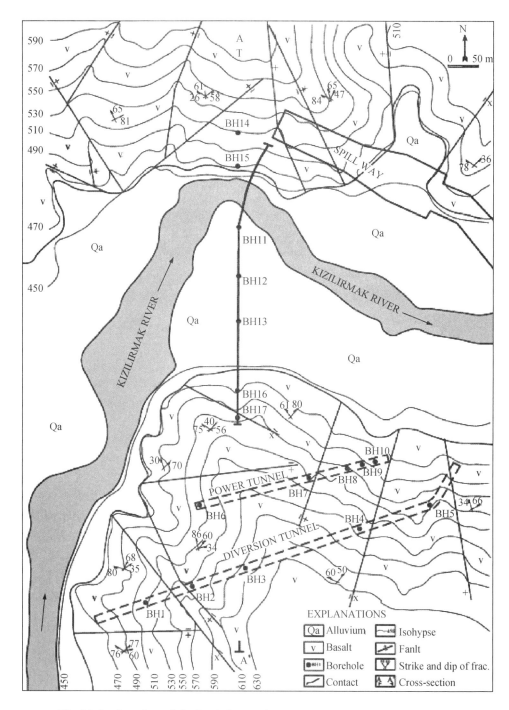

Fig. 22-2 Locations of the boreholes, main structures, and engineering geological map of the Obruk dam site.

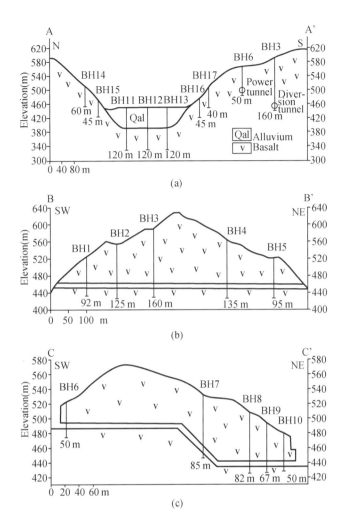

Fig. 22-3   Engineering geological cross-sections of the (a) dam axis,
(b) diversion tunnel and (c) power tunnel.

(Cited from Kocbay A, Kilic R. Engineering geological assessment of the Obruk dam site (Corum, Turkey) [J]. Engineering geology, 2006, 87(3-4): 141-148.)

New Words and Expressions

weathering	*n.*	风化
comprise	*v.*	包含
excavatibility	*n.*	可挖掘性
grouting	*n.*	灌浆
optimal	*adj.*	最佳的
core drilling in-situ testing		岩芯钻探原位测试

basalt	*n.*	玄武岩
irrigation	*n.*	灌溉
thalweg	*n.*	河流谷底线，深泓线
coefficient	*n.*	系数
dip direction		倾角
joint	*n.*	节理
outcrop	*n.*	露头，出露地表的岩层
volcanite	*n.*	火山岩
flysch	*n.*	［地质］复理层，复理石
interbed	*n.*	夹层
vicinity	*n.*	附近
lithology	*n.*	岩石学
plagioclase	*n.*	斜长石
coarse-grained	*adj.*	粗粒度的
pyroxene	*n.*	辉石
olivine	*n.*	橄榄石
phenocryst	*n.*	斑晶
opaque	*adj.*	不透明的
pyrite	*n.*	黄铁矿
chloritization	*n.*	绿泥石化作用，亚氯酸化作用
silicification	*n.*	硅化作用
argillization	*n.*	泥化，黏土化作用
gypsum	*n.*	石膏
Quaternary alluvium		第四纪冲积层

译文：Obruk 坝址的工程地质评估

介绍

决定岩石力学性质的主要因素是地质构造、矿物组成、不连续性和风化程度。因此，在施工前，工程技术人员应在现场和实验室预先确定主要工程结构物（如大坝）的基底岩石的地质特征和力学性质。从工程优化设计的角度来看，轴线位置，泄洪道、引水隧洞和电力隧洞穿越岩层的可开挖性，注浆条件，以及隧道支护系统的确定也具有重要的意义。工程地质勘察和岩石力学性质的研究主要包括不连续性勘测和岩芯钻探原位测试（Özsan and Akın，2002）。

本研究从工程地质角度调查了 Obruk 大坝基底的代表性玄武岩。该大坝建于 Anatolia 中部的 Çorum 西北约 50 公里的 Kızılırmak 河上（图 22-1）。该坝是为了水力发电和农业灌溉而修建的土石坝。大坝位于深泓线以上 67 米的位置,从底部到顶部的高度为 126 米,容积为 12 000 hm³,规划的发电功率为 203 MW,年发电量为 473 GWh。在 1979 年和 1983 年间,在 Lower Kızılırmak 项目的框架内,国家水利工程总局(DSI)根据 Obruk 大坝初始轴线区域的钻探工作完成了工程地质规划报告(DSI,1979,1983)。Kılıç(1999)研究了在规划的初始轴线区域内的玄武岩的地质力学特性,并根据风化程度将这些岩石分为五组。Koçbay(2003)对 Osmancık-Çorum 地区的玄武岩的风化特征和风化程度进行了研究,提出了一个与分解作用相关的分类方法。此外,Koçbay 和 Kılıç(2003)研究了这些玄武岩作为天然建筑材料的可行性。

因为在 Obruk 大坝的初始轴线区域的上游大约 500 米处有活动滑坡,所以技术人员重新选择了大坝的坝址。本研究主要涉及了新坝址的轴线区域的工程地质条件。为了研究玄武岩的性质在纵向和横向上的变化,我们采集了岩芯样品并进行 Lugeon 试验以确定水力传导系数和地下水位。我们利用了 17 个由国家水利工程总局(DSI)在大坝项目建设场地内打的钻孔。这 17 个钻孔的深度在 40 米至 160 米之间,总深度为 1 512 米。此外,我们确定了玄武岩中节理的主要走向和倾角,并从露头处采集了具有代表性的块体样品。

在初步确定了样品的矿物学、岩石学和化学特征后,我们根据风化程度对样品进行了分类。随后,我们确定了 172 个样品的物理性质和地质力学性质。利用这些数据,我们还绘制了工程地质图,以及 Obruk 大坝新轴线位置和电力、引水隧道位置的剖面图。在所有这些岩石特征的基础上,我们从隧道开挖的角度对岩体进行了分类。

坝址的地质情况

Lutetian Bayat 岩层在研究区域内有出露地表,它由火山岩和复理石夹层构成(MTA,1975)。在坝址附近普遍分布着 Bayat 地层,其主要岩性为玄武岩。玄武岩由细粒黑色基质,中等尺寸的斜长石晶体,相对粗粒的辉石和橄榄石斑晶,以及不透明的矿物组成。在局部,有黄铁矿存在于不透明的矿物相中。绿泥石化作用在玻璃质石粉和辉石矿物中广泛存在,而斜长石则转变成了粘土矿物。随着热液蚀变程度的增加,碳化、硅化、绿泥石化和泥化作用增强。这些作用沿着不连续面变得更加强烈,并且在裂缝中发生了石膏填充现象。在沉积层序的顶部分布的是第四纪冲积层,它由块体、砾石、砂子、淤泥和粘土材料组成(图 22-2)。

该研究区域距离土耳其最重要的地质构造之一——北安那托利亚断裂带(NAFZ)约 35 公里,因此受到了构造活动的强烈影响。在大坝轴线的左右位置出现了大量不同规模的东北—西南和西北—东南走向的正断层。构造作用和节理发育的影响也非常显著。图 22-2 和图 22-3 分别给出了坝址、引水及电力隧洞、溢洪道、左右边坡的工程地质图及剖面图。

Text 23　Engineering Geology — A Fifty Year Perspective

Introduction

Recently, the Journal, Engineering Geology, celebrated its 50th anniversary. Engineering Geology (referred to hereinafter as "the Journal") was founded in 1965 with the inaugural issue published in August of that year. More than 3,400 papers have been published in the Journal since then. To help celebrate the 50th anniversary of the Journal, a virtual special issue (VSI) that consists of thirty selected papers was published in 2015 (http://www.journals.elsevier.com/engineeringgeology/). These papers were selected by the two Chief Editors, Carlos Carranza-Torres and Charng Hsein Juang, with the assistance from the publisher, Kate Hibbert. The selection criteria included citations, the balance of the subjects, and contributions of individuals to the Journal. The preface of the VSI was written by Charng Hsein Juang with contributions from selected members of the Editorial Board of the Journal. This short communication, an enlarged version of the preface of the VSI, aims to summarize the history of the Journal, to discuss the future challenges faced by the engineering geology communities, and to provide new horizons for the young practitioners and researchers in this field.

Background and evolution of engineering geology

Engineering geology is a multidisciplinary subject of study at the intersection of earth sciences and engineering (particularly geological, civil, and mining engineering). Engineering geologists generally are trained as geologists, and they commonly have a background focused on the geologic and environmental factors that affect engineering design and construction. Their expertise also requires knowledge in soil and rock mechanics, groundwater, and surface water hydrology. The role of an engineering geologist is to understand the complexities of natural phenomena and geologic materials, and to describe them in a way that is readily usable in an engineering project.

Engineering Geology is an international research journal that is dedicated to serve the researchers and professionals working in the broad field of engineering geology. Indeed, the editorial of that inaugural issue clearly stated that the aim of the Journal was "to approach engineering from the geological angle, and to fill the gap between engineering and geology and to stimulate the publication of papers containing a significant content of both fields". The Journal published more than 3,400 papers during

20

the period of 1965-2015. The number of published papers steadily increased for the past 50 years (Fig. 23 - 1). The early theme of the Journal to emphasize "the need for engineers and geologists to communicate with each other" has greatly expanded over the past five decades to cover emerging technological and socioeconomic issues such as natural hazards, environmental concerns, and safety.

The papers published during the past 50 years can be classified into seven subject categories, as summarized in Table 23-1.

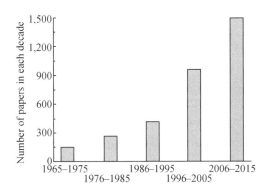

Fig. 23-1   Number of papers published in the Journal over the past 50 years

Table 23-1   Number of papers published in each subject category during the past 50 years

	1965-1975	1976-1985	1986-1995	1996-2005	2006-2015
PMPSR	74	87	131	226	473
EGEP	28	54	120	140	121
GEGE	18	54	26	119	91
SC	2	9	10	58	177
LS	13	24	44	165	425
GH	11	24	36	95	123
GEGR	4	11	45	155	82
Total	150	263	412	958	1,492
Unclassified	8	23	32	52	75

1) Physical and mechanical properties of soil, rock and rock masses (PMPSR); topics on treated soils, permafrost, and loess are also included in this category.

2) Engineering geology for engineering projects (EGEP); topics on hydropower stations, dams, and tunnels are also included in this category.

3) Geological engineering and geotechnical engineering (GEGE); topics on empirical or theoretical design of foundations, slopes, tunnels and related technologies are included in this category.

4) Site characterization (SC): topics on geophysical prospecting and remote sensing applications are also included in this category.

5) Landslides (LS): slope stability and slope erosion are also included in this category.

6) Geohazards other than landslides (GH): fault and earthquake, liquefaction, karst and sinkhole, and land subsidence are included in this category.

7) Geo-environment and geo-resources (GEGR): topics on radioactive wastes disposal, carbon dioxide capture and sequestration, waste landfilling, groundwater pollution, and development of groundwater resources are included in this category.

PMPSR is the most dominant category of papers published in the Journal; almost half of the papers published between 1965 and 1975 fall in this category. Over the past 50 years, the average of PMPSR has remained quite steady and accounts for approximately 30% of all papers (Fig. 23‑2). The papers categorized into EGEP account for approximately 14%. In 1995, the EGEP proportion reached the peak (30%) but has dropped quickly since then. After reaching a peak of approximately 20% in 1985, the GEGE category decreased significantly and averaged approximately 10% over the past 50 years. Site characterization (SC) is generally a part of the engineering project and thus this category can be difficult to differentiate from the aforementioned categories. However, this category has an increasing trend, reaching approximately 10% in the last 10 years. The Journal published numerous papers in the fields of the geo-hazard, geo-resources, and geo-environment. The landslide (LS) category is separated from the geo-hazard (GH) category because of its large proportion in the database. The long-term average proportion of LS papers is approximately 20%, but in the last ten years we have more than 400 LS papers, a substantial increase in both numbers and proportion of

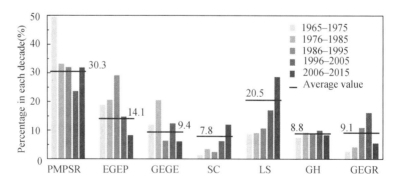

Fig. 23‑2 Variation of different subject categories published in the past five decades. PMPSR: Physical and mechanical properties of soil, rock and rock masses; EGEP: Engineering geology for engineering projects; GEGE: Geological engineering and geotechnical engineering; SC: Site characterization; LS: Landslides; GH: Geohazards; GEGR: Geo-environment and geo-resources

published papers. LS papers account for nearly one-third of all papers published in the Journal from 2006 to 2015. The number of geo-hazards (GH) papers was quite substantial, averaging approximately 10%. The papers in the geo-resources and geo-environment category (GRGE) also accounted for approximately 10%, with about half of the papers in this category being related to the subject of nuclear waste disposal.

Analysis of the subject preference of the published papers in each decade

1965 – 1975: In the early 1960's, numerous papers focused on the engineering behavior of problematic soils, such as quick clay or loess. Half of the papers published during this decade belonged to PMPSR category (Fig. 23 – 3). New methods were developed to improve the properties of these problematic soils. The first paper using remote sensing technology also appeared in this period, which used aerial photography to investigate the occurrence of laterites in northern Nigeria. It is interesting to note that induced seismicity was discussed extensively in the Journal during the middle of 1970's, and this is becoming a hot topic again now because of anomalous seismic activity recorded at some sites that could be related to the intense production of oil and gas shale, CO_2 sequestration, or geothermal energy extraction from hot dry rocks.

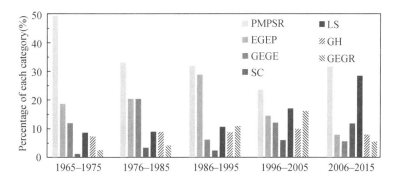

Fig. 23-3　Distribution of subject categories (i.e., percentage of each category) in each decade. PMPSR: Physical and mechanical properties of soil, rock and rock masses; EGEP: Engineering geology for engineering projects; GEGE: Geological engineering and geotechnical engineering; SC: Site characterization; LS: Landslides; GH: Geohazards; GEGR: Geo-environment and geo-resources

1976-1985: The papers belonging to PMPSR category dropped to less than one-third of the papers published in the Journal, whereas the papers related to EGEP and GEGE categories increased quickly during this period. Many papers in EGEP and GEGE from this decade were related to seismicity and reservoir projects. At the same time, the number of studies focusing on frost heave and ground freezing increased. Studies related to waste repositories first appeared in this decade.

1986-1995: During this period, the number of papers related to EGEP category reached a peak (~ 30%). This was probably related to the increasing budget for infrastructure development all around the world. Meanwhile, many other papers focused on the investigation of dam failures. The study of waste repositories also increased quickly in this decade, as researchers devoted much effort to environmental issues. As a result, the number of papers in the GEGR category increased significantly (greater than 10%).

1996–2005: In this decade the seven subject categories attracted similar attention from the researchers although the number of papers categorized into PMPSR was still greater than others and those categorized into SC was slightly less. The number of papers focusing on geo-environment and geo-resource issues reached a peak during this decade. Many LS papers appeared after 2000, which became a hot topic as reflected by the high citation counts. The number of papers related to earthquakes and active faults also increased in this decade.

2006–2015: The number of papers in the LS and SC categories increased quickly in this decade. In particular, the large magnitude 1999 Chi-Chi and 2008 Wenchuan earthquakes were two key events, which resulted in the extensive research on seismically induced landsliding and post-seismic slope hazards. However, the papers related to geoenvironment and geo-resource (GEGR) issues dropped during this period. Interestingly, the number of papers in the PMPSR category increased in this decade but those in EGEP and GEGE categories declined. The world-wide economic recession from 2008 might have contributed to the decline of infrastructure development and thus the number of papers in these two subject categories.

The interests of the engineering geology communities will undoubtedly continue to evolve to address the emerging needs in a changing world. Accordingly, the Journal will continue to adapt to better serve the professionals and researchers working in the broad field of engineering geology.

Opportunities and concerns in engineering geology

Three concepts clearly make engineering geology relevant in a changing world. First, burgeoning population growth clearly requires the construction of more infrastructure such as buildings, roads, airports, ports, and dams, which, in turn, will increase the demand for finding suitable sites or routes to meet these needs. These infrastructure needs will demand increasingly more quality services (e.g., site selection, environmental studies, and design) from the engineering geology professionals and more advanced research tools. Second, the increased need for industrial raw materials and ore minerals for building infrastructure and other uses (e.g., mining-related) will also enhance the important role of engineering geology. Third, the increase of losses from natural disasters, some of which could be exacerbated by land use and climate changes, will place a renewed focus on the relevance of engineering geology. Consequently, the future of engineering geology is very bright indeed, as opportunities are abundant for engineering geologists to be great contributors in making the world a better and safer place to live.

Two serious concerns affect this vision, however. First, the decrease in research revenues to fund new engineering geology research, coupled with the decrease in the number of engineering geology programs in higher education in the Western world, is detrimental to the scientific and technological advances of this discipline, which, in turn, adversely affects the education of young professionals. Second, the decrease in opportunity and quality of education and research will eventually affect the quality of engineering geology solutions in critical projects. Both problems can lead to a decrease in the relevance of engineering geology. To this end, engineering geology communities must reach out to policymakers and lawmakers because the ultimate funding to meet these future engineering geology challenges will come from them.

(Cited from Juang C H, Carranza-Torres C, Crosta G, et al. Engineering geology-A fifty year perspective [J]. Engineering Geology, 2016, 201: 67-70.)

New Words and Expressions

inaugural	*adj.*	开始的;开幕的
criteria	*n.*	标准,条件
practitioner	*n.*	从业者
multidisciplinary	*adj.*	各种学问的,多学科的
expertise	*n.*	专门知识,专门技术
hydrology	*n.*	水文学,水文地理学
dedicate	*n.*	致力,献身
stimulate	*vt.*	刺激,鼓舞,激励
technological	*adj.*	技术的,工艺的
socioeconomic	*adj.*	社会经济学的
natural hazards		自然灾害
permafrost	*n.*	永久冻土,永久冻结带
loess	*n.*	黄土
empirical	*adj.*	经验主义的
geophysical prospecting		地球物理勘探
remote sensing		遥感技术

land subsidence		地面沉降
sequestration	*n.*	隔离
laterite	*n.*	红土,铁矾土
anomalous	*adj.*	异常的,不规则的
gas shale		页岩气
geothermal energy		地热能
reservoir	*n.*	水库,蓄水池

译文:"工程地质"期刊——五十年的回顾

引言

最近,《Engineering geology》期刊(下文称"期刊")庆祝了创刊五十周年。《Engineering geology》创刊于1965年,其初刊于当年八月出版。自此以来已经有超过3400篇论文在这份期刊上发表。为了庆祝刊物创刊五十周年,一份由三十篇经过挑选的论文组成的虚拟特刊在2015年出版了。这些论文的遴选工作由两位主编Carlos Carranza-Torres和Charng Hsein Juang负责,并在出版商Kate Hibbert的协助下完成。论文挑选的标准包括其引用情况、主题的平衡性以及该论文对于期刊的贡献程度。这份虚拟特刊的前言由Charng Hsein Juang撰写,其中也包含了期刊编辑部的部分成员所作的贡献。作为虚拟特刊引言的扩展版本,这篇简短通讯的目的是总结《Engineering geology》期刊的历史,讨论工程地质学界在未来面临的挑战,并为该领域的年轻从业者及研究人员提供新的视野。

《Engineering geology》期刊的背景及演变

工程地质学是关于地球科学和工程学(尤其是地质工程、土木工程和采矿工程)的交叉学科。工程地质学家通常接受过地质学方面的专业培训,具备的知识背景使他们有能力研究地质及环境因素对工程设计与工程建设的影响。此外他们的专业技能还需要土力学及岩石力学、地下水及地表水文学方面的知识。一名工程地质学家的任务是去理解自然现象以及地质材料的复杂性,并以一种便于工程应用的方式将它们表述出来。

《Engineering geology》期刊是一份研究型的国际期刊,其致力于向那些在工程地质学广阔领域内从业的专家学者们提供服务。实际上,初刊在编辑时就明确了期刊创立的主旨在于"从地质学的角度走近工程,填补工程学与地质学之间的鸿沟,促进含有这两个领域内重要内容的论文的发表"。从1965年到2015年,这份期刊已经发表了超过3400篇论文,并且发表论文的数量在过去50年间稳步增长(图23-1)。期刊在早期强调的"工程师和地质学家互相交流的必要性"这个主旨在过去五十年间也扩展至新兴的技术问题与社会经济问题,比如自然灾害、环境问题和安全问题。

在过去50年中发表的论文可以被分为以下七种类型,如表23-1所示:

1)土体、岩石及岩体的物理力学性质(以下简称PMPSR);处理土、永冻土以及黄土等主题也包括在这一类中。

2) 工程项目涉及的工程地质学问题(以下简称 EGEP);水电站、水坝以及隧道等主题也包括在这一类中。

3) 地质工程与岩土工程(以下简称 GEGE);基础、边坡、隧道以及相关工艺的经验与理论设计等主题也包括在这一类中。

4) 场地特征(以下简称 SC);地球物理勘探和遥感技术的应用等主题也包括在这一类中。

5) 滑坡崩塌(以下简称 LS);边坡稳定性与边坡侵蚀等主题也包括在这一类中。

6) 除滑坡崩塌以外的地质灾害(以下简称 GH);此类别包括地层断裂与地震、砂土液化、喀斯特效应与落水洞以及地面沉降等主题。

7) 地质环境与地质资源(以下简称 GEGR);此类别包括放射性废物的处置、二氧化碳的捕捉与封存、垃圾填埋、地下水污染以及地下水资源发展等主题。

PMPSR 是期刊发表的论文中最主要的一类——在 1965—1975 年所发表的论文中有将近一半属于这一类。在过去的 50 年间,PMPSR 的平均论文数量保持得相当稳定,约占总论文数量的 30%(图 23-2)。而被分至 EGEP 大类中的论文数量则大约占了总论文数的 14%。1995 年,EGEP 部分的论文比重达到了顶峰(30%),但此后又快速下降。GEGE 大类的论文在 1985 年达到了一次大约占比 20% 的顶峰之后,又出现了大幅度的缩减,其在 50 年间的平均占比约为 10%。通常情况下场地特征(SC)是工程项目的一部分,因此将这一类论文从前述类别的论文中区分出来是很困难的。但是此类别的论文也有着增长的趋势,在最近 10 年间占比达到了近 10%。这份期刊发表了大量与地质灾害、地质资源以及地质环境领域相关的论文。将滑坡崩塌类(LS)从地质灾害类别(GH)中分离出来是因为它在数据库中占有很大的比例。滑坡崩塌类(LS)论文长期以来的平均占比约为 20%。在最近的 10 年间我们已经发表了超过 400 篇的相关论文,无论在数量上还是总体占比上都有着很大的提升。滑坡崩塌类(LS)论文在 2006—2015 年发表的论文中占了近三分之一。地质灾害类(GH)论文的数量也很大,平均占比约为 10%。地质资源与地质环境类(GRGE)的论文也占到了将近 10%,其中大约一半的论文与核废料处置的主题有关。

关于已发表论文在每一个十年内的主题倾向分析

1965—1975:在 20 世纪 60 年代早期,大量的论文关注问题土(problematic soils)的工程性质,如超灵敏粘土或黄土。这十年间发表的论文有一半属于 PMPSR 范畴(图 23-3)。人们为了改善这些问题土的性质提出了一些新的方法。第一篇关于遥感技术运用的论文也出现在这一时期,其利用航空摄影技术调查了尼日利亚北部红土的产生情况。有趣的是,诱发地震问题在七十年代中期的期刊上曾被广泛讨论,而最近再一次成为了热门话题。这可能与石油和页岩气的大量开采、二氧化碳封存或干热岩体内的地热能源开采有关的场地异常性地震活动记录有关。

1976—1985:PMPSR 类的论文数量下降到占期刊总发表论文数量的三分之一以下,而 EGEP 和 GEGE 类的相关论文在此期间又迅速增长。这十年间,EGEP 和 GEGE 类的许多论文都与地震活动及水库项目有关。与此同时,关注冻胀和地面冻结的研究数量有所增加。最早的与废物储存库相关的研究出现在这十年间。

1986—1995:在此期间,EGEP类相关论文的数量达到峰值(约30%)。这可能与全世界范围内基础设施建设的预算增长有关。与此同时,许多其他方面的论文则集中于对水坝失效问题的调查。由于研究人员在环境问题上投入了大量的精力,关于废物储存库的研究在这十年间也迅速增加。因此,GEGR类别的论文数量显著增加(超过10%)。

1996—2005:在这十年间,7个类别的主题吸引了学者们相似的关注度。其中PMPSR的论文数量仍然比其他分类的论文数量略多,而SC的论文数量则是略少。关注地质环境和地质问题的论文数量在这十年间达到了顶峰。在2000年后出现了许多LS类的论文,其很高的引用率反映出它成为了一个热门的研究课题。在这十年间与地震和活动断裂相关的论文数量也有所增加。

2006—2015:在这十年间,LS和SC类的论文数量迅速增加。尤其是1999年集集大地震和2008年汶川大地震成为了两个关键事件。它们引发了对地震诱发滑坡和震后边坡灾害治理的广泛研究。然而,与地质环境和地质资源(GEGR)相关的论文在这一时期内数量有所下降。有趣的是,PMPSR类别的论文数量在这十年间有所增加,而EGEP和GEGE类别的论文数量则有所下降。这可能是由于2008年以来的全球经济衰退造成了基础设施建设投入的紧缩,从而导致这两个类别的论文数量的下降。

毫无疑问,工程地质学界的研究方向将继续发展变化以满足在这个不断变化的世界中出现的新兴需求。因此,本期刊将继续适应这种发展变化,从而更好地为在工程地质学广阔领域内从业的专家和学者们提供服务。

工程地质学的机遇与忧虑

以下三个概念清楚地说明了工程地质学对于不断变化的世界具有重要的作用。第一,迅速增长的人口显然需要人类建设更多的基础设施,如建筑、道路、机场、港口和水坝等。这就需要人们去寻找更多合适的场地或路线以建设这些基础设施。工程地质专家及其采用的更先进的研究工具需要不断提供更高质量的服务(如选址、环境研究和设计)以满足基础设施建设的需求。第二,为了满足基础设施建设以及其他一些行业(例如与采矿相关)的需要,人们对工业原料和矿石矿物的需求也在不断增长,这也将增强工程地质学的重要性。第三,自然灾害造成的损失正在增加,且其中一些损失还可能因为土地利用和气候变化而加剧,这促使人们重新关注工程地质学的重要性。因此,工程地质学的前景的确非常光明。工程地质学家有丰富的机会为实现"让世界成为一个更美好和更安全的居住地"这一目标作出巨大的贡献。

然而,有两个严重的担忧影响了上述看法。首先,可用于资助新的工程地质研究的财政收入在不断紧缩,而且西方国家高等教育中工程地质专业的数量亦在减小。这些都会对这门学科在科学技术方面的进步与发展产生不利影响,进而也会对青年专业人才的教育造成负面影响。其次,在教育和科研方面的机会减少和质量下降最终会影响重大项目的工程地质解决方案的质量。这两个问题都有可能降低工程地质学的重要性。为此,工程地质学界必须与决策者及立法者接触,因为工程地质在面对未来挑战时所需要的资金最终还是来自他们。

Text 24   The Collapse of the Sella Zerbino Gravity Dam

The failure of the Sella Zerbino dam

After a long dry period, at 6:15 a.m. of 13th August 1935 an exceptionally severe rainfall storm hit the Orba basin (Natale and Petaccia, 2013). At 7:00 a.m. the rain intensity increased and kept on without interruptions until 3:00 p.m. The rain reached his highest intensity between 7:00 and 8:00 a.m. and between 2:00 p.m. and 3:00 p.m. At 9:30 a.m. the siphon outlets operated to their full capacity (Petaccia and Fenocchi, 2015). At 10:45 a.m. the water began to enter the side spillway and soon afterwards the flood inflow exceeded the maximum capacity of the spillways. After 15 min the bell valve was clogged up with sediments and debris and there was no way to re-open it. At 12:30 a.m. both of the dams were overtopped. At 1:15 p.m. the Sella Zerbino dam abruptly collapsed (Alfieri, 1936). Historical witnesses confirm that the collapse started from the left side of the dam. The remaining parts of the dam failed sequentially. On August 15, 1935 a new flood wave swept down the wreckage of the left abutment. The dam failure flooded an area of almost 70 km^2 and caused 111 casualties, 97 of which in Ovada town (Natale et al., 2008; Petaccia et al., in press).

Geologic and geotechnical site characterization

The geological reports prepared for the design of the two Zerbino dams were mainly referred to the location of the Bric Zerbino dam (Lelli, 1937; Peretti, 1937). Recent surveys on Sella Zerbino site (Capponi, 2014; Bonaria and Tosatti, 2013) highlighted the extremely poor quality of the bedrock. Fig. 24-1 shows the transversal cross section at the dam site.

The mechanical characteristics of the bedrock were retrieved from the results of the experimental investigation campaign funded in 1980 by Piedmont Regional Administration to assess the budget costs of a multipurpose plant aimed to revitalize the exploitation of the Orba River. The investigation included a series of geophysical refraction surveys (Calvino and Siccardi, 1980) from which a total of 15 subsoil profiles were constructed. The average low-strain elastic moduli of the undisturbed rock formations were then estimated from the measured speed of propagation V_P of compressional waves. The values of V_P typical of a compact rock could only be found below 20 m from the free surface. Three boreholes with continuous sampling were drilled to reconstruct the lithostratigraphic profile along the axis of the Sella Zerbino dam (Fig. 24-2). The rock formations were distinguished based on the RQD (Rock Quality Designation) geomechanical parameter.

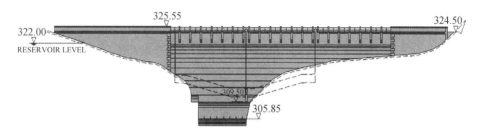

Fig. 24-1　Downstream view of Sella Zerbino dam 13/8/1935

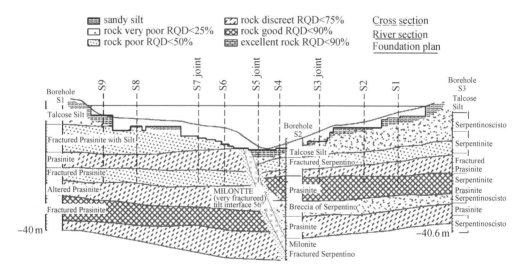

Fig. 24-2　Longitudinal vertical cross-section at the site of foundation of the Sella Zerbino dam (view from upstream)

The hydraulic conductivity of the geologic formations was measured through standard in situ Lugeon tests. The subsoil at the foundation site of the dam is locally fractured; therefore a distinction was made between primary and secondary permeability. Furthermore, different mechanical properties of the various lithotypes produce systems of discontinuities with different hydraulic parameters. Fig. 24-2 shows qualitatively that the bedrock is highly fractured and inadequate to bear the stresses induced by the weight of the Sella Zerbino dam. As mentioned above, this was noted during the construction works when the grout curtain was injected.

The top soil layer is a sandy silt originated from weathering of the underlying bedrock and a highly fractured Serpentinite with talcose silt; its thickness varies from 1 to 4 m below the ground level. A large variability of geomechanical characteristics characterizes the intermediate geologic formation at the right abutment (borehole S3 in Fig. 24-2). This is also caused by the presence of a mylonite interface in the interior of the riverbed. In the central part of the dam cross-section the intermediate geologic formation has relatively homogeneous properties (borehole S2 in Fig. 24-2). However, the terminal part of the borehole intercepts the mylonite with some visible Serpentinite pieces after crossing some degraded Serpentinite breccia.

We tested 3 concrete samples of the remains of Sella Zerbino dam recovered during one of our on-site surveys; the results are reported in Table 24-1. In this table R_{ck} denotes the average cubic strength, f_{ck} the average cylindrical strength, f_{ctm} the average tensile strength and F_{cm} the average elastic Young modulus of the concrete.

Table 24-1   Mechanical characteristics of the cyclopean concrete of the Sella Zerbino dam calculated according to the formulas of the current Italian Building Code (NTC08)

Cubic samples	R_{ck}	f_{ck}	f_{ctm}	f_{cm}	E_{cm}
	$[N/mm^2]$	$[N/mm^2]$	$[N/mm^2]$	$[N/mm^2]$	$[N/mm^2]$
1	37.20	30.88	2.95	38.88	33,062.50
2	28.87	23.96	2.49	31.96	31,174.51
3	30.17	25.04	2.57	33.04	31,486.59
Average	32.08	26.63	2.67	34.63	31,907.87

Possible causes of the collapse of Sella Zerbino dam

Following the collapse of the Sella Zerbino dam, different causes were blamed for the disaster (Accusani, 1936; De Marchi, 1937; Mangiagalli, 1937) including:

1. Failure of the internal body of the barrage due to inadequate concrete resistance.

2. Instability due to the uplift force. The design did not consider this force, even though it was required by the technical Standard of that time (Ministry Rule No. 1309 of April 2, 1921).

3. Instability due to the scour caused by the plunging of the water overtopping the dam.

Since both the trial report and our preliminary analyses verified that the cause in 1 did not occur, the following sections thoroughly investigate causes 2 and 3.

(Cited from Petaccia G, Lai C G, Milazzo C, et al. The collapse of the Sella Zerbino gravity dam [J]. Engineering geology, 2016, 211: 39-49.)

New Words and Expressions

siphon	*n.*	虹吸管
spillway	*n.*	溢洪道,泄洪道
clogged	*adj.*	阻塞的,堵住的
debris	*n.*	碎片,残骸
abutment	*n.*	临界,接界
retrieve	*n.*	恢复,重新取回
revitalize	*vt.*	使恢复元气
lithostratigraphic	*adj.*	岩石层位学的
formations	*n.*	形成,构造
conductivity	*n.*	导电性,传导性
fractured	*adj.*	断裂的
grout curtain		灌浆帷幕
abatement	*n.*	减少,减轻,消除
mylonite	*n.*	糜棱岩
intercept	*vt.*	拦截,截断
serpentinite	*n.*	蛇纹岩
breccia	*n.*	角砾岩
barrage	*n.*	拦河坝
plunge	*v.*	急降
preliminary	*adj.*	初步的,开始的,预备的

译文:塞拉-泽尔比诺重力坝的坍塌

塞拉-泽尔比诺水坝的破坏

在经历了一段较长的干旱期后,1935 年 8 月 13 日上午 6:15,一场异常严重的暴雨袭击了奥巴盆地(Natale and Petaccia, 2013)。早上 7:00,雨量开始增加并一直持续到下午 3:00。雨量在上午 7:00 至 8:00 和下午 2:00 至 3:00 之间达到最高强度。上午 9:30,虹吸式出水流道的运转达到满负荷状态(Petaccia and Fenocchi, 2015)。上午 10:45,雨水开始进入侧部泄洪道,之后不久洪水流入量超过了泄洪道的最大容量。15 分钟后,钟型阀被沉积物和碎屑堵塞而无法再次打开。上午 12:30,两座大坝都被洪水淹没。下午 1:15,塞拉-泽尔比诺大坝突然发生坍塌(Alfieri, 1936)。目击者证实坍塌是从大坝的左侧开始发生,此后大坝的其余部分陆续破坏。1935 年 8 月 15 日,一股新的洪峰席卷了左岸坝肩的残骸。溃坝引发的洪水淹没了近 70 平方公里的区域,造成了 111 人死亡,其中 97 人是奥瓦达镇的居民

(Natale., 2008；Petaccia et al., in press)。

地质和岩土场地特征

工程技术人员在设计两座泽尔比诺大坝时使用的地质报告主要参考了布里奇-泽尔比诺大坝所在位置的地质条件(Lelli, 1937；Peretti, 1937)。最近对塞拉-泽尔比诺场地进行的调查结果(Capponi, 2014；Bonaria Tosatti, 2013)突显了它极差的基岩质量。图 24-1 为坝址处的横截面示意图。

基岩的力学特性是从一次试验勘察的结果中获取的。该勘察活动由皮埃蒙特区管理局在 1980 年资助完成,其目的是评估一个为了振兴奥尔巴河而开发的多用途设施的预算成本。勘察工作包括一系列地球物理折射勘测(Calvino and Siccardi, 1980),其中涉及 15 个地基剖面。根据实测压缩波的传播速度 V_P,可以估算出原状岩层的平均低应变弹性模量。致密岩石的 V_P 值只能在自由地表面以下 20 m 深度处获取。在塞拉-泽尔比诺大坝的轴线上钻了三个连续取样的钻孔以重新绘制岩石地层剖面(图 24-2)。技术人员基于地质力学参数 RQD(岩石质量指标)来区分不同的岩层。

技术人员还通过标准现场透水试验测量了地层的渗透系数。由于坝基的地基土局部破碎,技术人员对原生渗透率和次生渗透率进行了区分。此外,各种类型的岩石具有不同的力学性质,因此形成了具有不同水力参数的不连续系统。图 24-2 定性地显示出大坝的基岩严重破碎,不足以承受塞拉-泽尔比诺大坝的自重应力。如前所述,技术人员在灌浆帷幕的施工过程中注意到了这一点。

表层土由下伏基岩风化而来的砂质粉土和高度破碎的蛇纹岩及滑石粉土组成,其厚度在地面以下 1 至 4 米不等。右坝肩位置的中间地层的地质力学特征表现出巨大的变异性(图 24-2 中的钻孔 S3)。这种变异性也是由于在河床内部存在糜棱岩的接触面而形成的。在大坝横截面的中心部位,中间地层具有相对均匀的性质(图 24-2 中的钻孔 S2)。然而,钻孔在穿过一些退化的蛇纹石角砾岩之后,钻孔的末端部分碰到了含有可见蛇纹岩片的糜棱岩。

我们对塞拉泽尔比诺大坝遗址的 3 个混凝土试样进行了测试。这些混凝土试样是我们在一次现场调查中回收的。测试结果见表 24-1。在本表中 R_{ck} 表示混凝土立方体强度的平均值,f_{ck} 表示圆柱体强度的平均值,f_{ctm} 表示抗拉强度的平均值,E_{cm} 表示杨氏弹性模量的平均值。

塞拉-泽尔比诺水坝垮塌的可能原因

在塞拉-泽尔比诺大坝坍塌之后,灾难的发生被归咎于不同的原因(Accusani, 1936；De Marchi, 1937；Mangiagalli, 1937),其中包括:

1. 混凝土抗力不足导致拦河坝内部的失效破坏。

2. 抬升力造成的不稳定性。尽管当时的技术标准对此有所要求,但设计中并没有考虑到这种力(Ministry Rule No. 1309 of April 2, 1921)。

3. 大坝溢流冲刷造成的不稳定性。

由于试验报告和我们的初步分析都证实了原因 1 并未发生,所以后续部分将对原因 2

和原因 3 进行深入的探究。

Text 25　Stability Analysis of Llerin Rockfill Dam

Introduction

Mining activities, especially industrial coal washeries, generate a series of waste materials that must be treated so as to mitigate their effect on the environment. Dams are usually employed to store tailings with a high-water content. The natural basins formed by the relief of the land (valleys) and a containment dam are exploited for this purpose. The aim of these tailings lagoons is twofold: to have a place to store a large amount of sludge and, at the same time, an area where this sludge is gradually decanted and compacted. This process is common practice in the Asturian mining valleys in Northern Spain, where the location of coal washeries, the proximity of built-up areas and the conditions of the relief of the land make this alternative the most feasible one from an economic point of view.

In this paper, we shall study how to analyze the stability of the Llerin rockfill dams used for this purpose. Although they are no longer operative, many of these dams require maintenance and monitoring tasks so as to guarantee their stability and endurance over time, especially when they are close to densely populated areas. Aside from the geological structure of the valley that is to be used for storage and its geotechnical characteristics, the behavior of the rockfill materials must also be ascertained by means of in situ testing with the aim of reducing the deviations produced in laboratory tests related to the scale factor of the samples (Maranha-das-Neves and Veiga-Pinto, 1988; Lewis, 1956; Douglas, 2002). One of the most important questions to arise in this analysis is the determination of the friction force between the materials that make up the rockfill, as well as the interface between the dam and the rock substrate on which it rests. Once the strength properties of the dam and the rock substrate are known, we are in a position to construct a model that allows us to estimate the safety factor of the rockfill dam and to take the necessary corrective measures to ensure its continuity over time.

To study these questions, Section 2 presents the shear criteria for the study of the rockfill dam, taking Barton—Kjaernsli's criterion as reference. Section 3 discusses the need to carry out in situ tests in order to ascertain the real behavior of the dam, for which purpose a direct shear test was designed. Section 4 presents the results obtained in the study of the Llerín tailings lagoon, together with a description of the geotechnical characteristics of the rockfill containment dam. The in situ test results are analyzed with the aim of defining the parameters of the Barton—Kjaernsli shear criterion. Finally, we present the results obtained using the SLOPE computer program to determine the dam's safety coefficient.

Shear criteria in rockfill dams

Traditionally, shear criteria for this type of material have been based on the fitting of the linear failure envelope of triaxial tests on small laboratory samples, known as the Mohr—Coulomb criterion. In this criterion, both the cohesion as well as the friction of the material are considered independently of their stress state. Moreover, in the case of rockfill type materials, cohesion is assumed to be null (purely frictional).

Barton and Kjaernsli developed a stress model for determining the shear strength of rockfill materials on the basis of the study of jointed rock masses (Barton and Kjaernsli, 1981), starting out from the premise that these constitute purely frictional materials (Barton and Choubey, 1977). They maintain that the internal friction angle of the rockfill may be determined from the simple compression strength of the original rock mass, the average particle size, the degree angularity and the porosity of the compacted material. These authors propose a parabolic fitting function between the confining pressures and the shear strength of the rockfill material. Thus, it is assumed that friction is not constant, but that it varies with the normal stress (σ_n) that acts on the shear plane according to the following expression:

$$\tau = \sigma_n \tan\left[R \log\left(\frac{S}{\sigma_n}\right) + \varphi_b\right] \tag{1}$$

where:

• R is the equivalent roughness of the system and depends on the "geological" origin of the material, the shape of the particles and the porosity of the rockfill.

• S is the equivalent strength of the granular system, which is a function of the simple compression strength of the rock (σ_c) and of the mean diameter of the fragments.

• φ_b is the basic friction angle.

We may estimate both the equivalent roughness parameter (R) and that of the equivalent strength (S) by means of abacuses (Barton and Kjaernsli, 1981). Charles and Watts, 1980 had previously been proposed this type of parabolic fitting for rockfill dams. Sarac and Popovic, 1985 studied this function for different materials, and Matsumoto and Watanabe, 1987 carried out the fitting by means of triaxial tests. The latter authors introduced the concept of "apparent cohesion", understood as a binding together, i.e. as the wedging that exists between the particles in the rockfill due to shear effects. Previously, Marsal and Resendiz, 1975 proposed a method for estimating the strength properties of rockfill material. These authors put forward a new test that consisted in subjecting three rockfill fragments to concentrated forces in their contacts similar to those that they have to support in reality. To do so, they simultaneously apply forces transmitted by means of a steel plate to three particles of approximately the same average size. The load that produces shearing of the first of the grains is measured and abacuses are designed on the basis of these data that allow us to determine the internal friction angle of the rockfill. Other works proposed conceptual models explaining the characteristic behavior of rockfill and its dependence on loads and water action (Alonso and Olivilla, 2005; Alonso, 2003; Oldecop and Alonso, 2003; Chen et al., 2007). Recently, Varadarajan et al., 2006 proposed a constitutive modelbased on the concept of "disturbed state", establishing a series of relationships between the constants of the material and the size of the particles that enables the behavior of the rockfill materials to be predicted. The modeled rockfill materials are subjected to drained triaxial tests using large size specimens (Varadarajan et al., 2006). Ramamurthy proposed a non-linear shear strength in which the effective cohesion and the effective angle of shearing cannot considered as constants (Ramamurthy, 2001).

Other authors have likewise proposed experimental devices to carry out in situ direct shear tests (Helgstedt et al., 1997; Cea-Azañedo and Olalla-Marañón, 1991) which enable the estimation of the friction angle under real conditions in which the rockfill material is to be found.

(Cited from Oyanguren P R, Nicieza C G, Fernández M I Á, et al. Stability analysis of Llerin Rockfill Dam: An in situ direct shear test [J]. Engineering Geology, 2008, 100(3-4): 120-130.)

New Words and Expressions

rockfill dam		堆石坝
tailing	n.	尾料,屑,残渣
lagoon	n.	潟湖
decant	v.	移入其他容器,轻轻倒出
The relief of the land		地形
in situ test		现场试验
substrate	n.	基质,基底
fitting	n.	拟合
premise	n.	前提

| parabolic | *adj*. | 抛物线的 |
| wedge | *n*. | 楔入 |

译文：列林堆石坝的稳定性分析

引言

采矿活动，尤其是工业洗煤厂会产生一系列的废料。对这些废料必须加以处理以减轻对环境的影响。大坝就经常被用来存储高含水量的尾矿。为了实现这一目的，可以利用现场地形(山谷)和围坝形成的天然盆地作为尾矿的存储场所。这些尾矿湖具有双重功能：一方面可以储存大量的工业淤渣，同时在尾矿湖中还可以逐步倾倒和压实这些工业淤渣。这在西班牙北部的阿斯图里亚采矿山谷中是一种非常普遍的做法。在那里，洗煤厂的位置、邻近建筑区域的情况以及地形条件使得这一替代方案从经济角度来看是最可行的。

列林堆石坝就是为此目的而建设的。本文将研究如何对列林堆石坝进行稳定性分析。虽然此大坝已不再运行，但许多类似的大坝仍需要维护及监测以确保大坝长期的稳定性和耐久性，尤其是当它们靠近人口稠密地区时则更为必要。除了存储区的河谷的地质结构及其岩土工程特性之外，技术人员还必须通过现场试验确定堆石材料的性能以减少在室内试验中由试样的尺寸效应造成的误差(Maranha-das-Neves and Veiga-Pinto, 1988；Lewis, 1956；Douglas, 2002)。在分析中出现的最重要的问题之一是如何确定堆石体的组成材料间的摩擦力以及大坝与基岩之间的接触面性质。一旦大坝和基岩的强度特性已知，我们就能够构建一个模型来估算堆石坝的安全系数，并采取必要的修正措施以确保其随时间的连续性。

为了研究这些问题，第2节以Barton-Kjaernsli准则为参考，提出了用于堆石坝研究的剪切准则。第3节讨论了通过现场试验确定大坝的实际工程性能的必要性，并为此设计了直接剪切试验。第4节介绍了列林尾矿湖的研究成果，并描述了堆石坝的岩土工程特性。我们还分析了现场试验结果以确定Barton-Kjaernsli剪切准则的参数。最后，我们利用SLOPE计算机程序确定了大坝的安全系数。

堆石坝的剪切准则

传统意义上，此类材料的剪切准则基于对由小型试样的三轴试验得到的线性破坏包络线的拟合，被称为莫尔—库仑准则。根据这一准则，材料的内聚力和摩擦力都被认为与应力状态无关。此外，对于堆石型材料，其内聚力假定为零(即纯摩擦)。

Barton和Kjaernsli在研究节理岩体的基础上建立了确定堆石材料抗剪强度的应力模型(Barton and Kjaernsli, 1981)，而其前提是这些材料为纯摩擦材料(Barton and Chou bey, 1977)。他们认为，堆石体的内摩擦角可由原始岩体的单轴抗压强度、平均粒径、颗粒的棱角角度和压实材料的孔隙率确定。两位学者提出了围压与堆石材料的抗剪强度之间的抛物线拟合函数。因此，这里假定摩擦力不是常数，而是随作用于剪切面上的法向应力(σ_n)而变

化,如下式所示:

$$\tau = \sigma_n \tan\left[R \log\left(\frac{S}{\sigma_n}\right) + \varphi_b \right] \tag{1}$$

其中:

- R 是系统的等效粗糙度,取决于材料的"地质"起源、颗粒的形状和堆石体的孔隙率。
- S 是颗粒体系的等效强度,它是岩石的单轴抗压强度(σ_c)及其碎屑平均直径的函数。
- φ_b 是基本摩擦角。

我们可以通过一些算法来估计等效粗糙度参数(R)和等效强度(S)(Barton and Kjaernsli,1981)。Charles 和 Watts 曾于 1980 年提出针对堆石坝的抛物线拟合函数。Sarac 和 Popovic 于 1985 年研究了不同材料的抛物线拟合函数,而 Matsumoto 和 Watanabe 于 1987 年通过三轴试验开展了拟合工作。后者引入了"表观内聚力"的概念,它可以理解为一种结合型式,即由于剪切作用而存在于堆石体颗粒之间的楔入现象。更早以前,Malsal 和 Resendiz 于 1975 年提出了一种估算堆石材料强度特性的方法。他们设计了一种新型试验,使三个堆石碎片在接触部位受到与它们实际承受的力相类似的集中力。为此,他们把钢板作为传递工具,向平均尺寸大致相同的三个颗粒同时施加力。然后,他们测量了当初始颗粒发生剪切时的荷载,并根据这些数据设计了可以确定堆石体的内摩擦角的算法。其他的研究工作提出了概念模型以解释堆石料的性能特征及其与荷载及水的作用的相关性(Alonso and Olivilla,2005;Alonso,2003;Oldecop and Alonso,2003;Chen et al.,2007)。最近,Varadarajan 等人于 2006 年提出了基于"扰动状态"概念的本构模型,建立了材料常数与颗粒尺寸之间的一系列关系,从而能够预测堆石材料的性能。他们运用大尺寸试样进行了堆石材料的排水三轴试验模拟(Varadarajan et al.,2006)。Ramamurthy 于 2001 年提出了一种非线性剪切强度准则,在这一准则中有效内聚力和有效剪切角度不是常数(Ramamurthy,2001)。

其他作者也提出了能够进行原位直剪试验的试验装置(Helgstedt et al.,1997;Cea-Azañedo and Olalla-Marañón,1991),从而能够在真实条件下估算堆石材料的摩擦角。

References

[1] Baker R, Frydman S. Unsaturated soil mechanics: Critical review of physical foundations [J]. Engineering Geology, 2009, 106(1-2): 26-39.

[2] Barnett T P, Pierce D W, Hidalgo H G, et al. Human-induced changes in the hydrology of the western United States [J]. Science, 2008, 319(5866): 1080-1083.

[3] Brooks B A. Seeing is believing [J]. Science, 2012, 338(6104): 207-208.

[4] Buffett B A. Taking earth's temperature [J]. Science, 2007, 315(5820): 1801-1802.

[5] Chai J C, Shen S L, Zhu H H, et al. Land subsidence due to groundwater drawdown in Shanghai [J]. Geotechnique, 2004, 54(2): 143-147.

[6] Comerio M C. Can buildings be made earthquake-safe? [J]. Science, 2006, 312(5771): 204-206.

[7] Deming D. Origin of the ocean and continents: a unified theory of the earth [J]. International Geology Review, 2002, 44(2): 137-152.

[8] Donal M R. Structural geology: An introduction to geometrical techniques [M]. 4th ed. Cambridge: Cambridge University Press, 2009.

[9] Ekström G, Stark C P. Simple scaling of catastrophic landslide dynamics [J]. Science, 2013, 339 (6126): 1416-1419.

[10] Geological development of an area [EB/OL]. https://en.wikipedia.org/wiki/Geology.

[11] Iverson R M, Reid M E, Iverson N R, et al. Acute sensitivity of landslide rates to initial soil porosity [J]. Science, 2000, 290(5491): 513-516.

[12] Juang C H, Carranza-Torres C, Crosta G, et al. Engineering geology-A fifty year perspective [J]. Engineering Geology, 2016, 201: 67-70.

[13] Keefer D K, Larsen M C. Assessing landslide hazards [J]. Science, 2007: 1136-1138.

[14] Kocbay A, Kilic R. Engineering geological assessment of the Obruk dam site (Corum, Turkey) [J]. Engineering geology, 2006, 87(3-4): 141-148.

[15] Long M D. How mountains get made [J]. Science, 2015, 349(6249): 687-688.

[16] Ojha C, Berndtsson R, Bhunya P. Engineering hydrology [M]. 1st ed.Oxford: Oxford University Press, 2008.

[17] Ostřihanský L. Causes of earthquakes and lithospheric plates movement [J]. Solid Earth Discussions, 2012, 4: 1411-1483.

[18] Oyanguren P R, Nicieza C G, Fernández M I Á, et al. Stability analysis of Llerin Rockfill Dam: An in situ direct shear test [J]. Engineering Geology, 2008, 100(3-4): 120-130.

[19] Pereira M F, Silva J B, Ribeiro C. The role of bedding in the formation of fault-fold structures, Portalegre-Esperança transpressional shear zone, SW Iberia [J]. Geological Journal, 2010, 45 (5-6): 521-535.

[20] Petaccia G, Lai C G, Milazzo C, et al. The collapse of the Sella Zerbino gravity dam [J]. Engineering geology, 2016, 211: 39-49.

[21] Phillips R J, Zuber M T, Solomon S C, et al. Ancient geodynamics and global-scale hydrology on Mars [J]. Science, 2001, 291(5513): 2587-2591.

[22] Reager J T, Gardner A S, Famiglietti J S, et al. A decade of sea level rise slowed by climate-driven hydrology [J]. Science, 2016, 351(6274): 699-703.

[23] Roy K K. Potential theory in applied geophysics [M]. 1st ed. Berlin Heidelberg: Springer Science & Business Media, 2008.

[24] Willenberg H, Loew S, Eberhardt E, et al. Internal structure and deformation of an unstable crystalline rock mass above Randa (Switzerland): Part I-Internal structure from integrated geological and geophysical investigations [J]. Engineering Geology, 2008, 101(1-2): 1-14.

[25] Zang A, Stephansson O, Heidbach O, et al. World stress map database as a resource for rock mechanics and rock engineering [J]. Geotechnical and Geological Engineering, 2012, 30(3): 625-646.